If I Only Changed the Software,
Why Is the Phone on Fire?

Embedded Debugging Methods Revealed

If I Only Changed the Software, Why Is the Phone on Fire?

Embedded Debugging Methods Revealed

Technical Mysteries for Engineers

by Lisa Simone

AMSTERDAM • BOSTON • HEIDELBERG • LONDON
NEW YORK • OXFORD • PARIS • SAN DIEGO
SAN FRANCISCO • SINGAPORE • SYDNEY • TOKYO

Newnes is an imprint of Elsevier

Newnes

Newnes is an imprint of Elsevier
30 Corporate Drive, Suite 400, Burlington, MA 01803, USA
Linacre House, Jordan Hill, Oxford OX2 8DP, UK

 Recognizing the importance of preserving what has been written, Elsevier prints its books on
acid-free paper whenever possible.

Library of Congress Cataloging-in-Publication Data

Simone, Lisa K.
  If I only changed the software, why is the phone on fire? : embedded
debugging methods revealed : technical mysteries for engineers / by Lisa
K. Simone.
       p. cm.
  Includes index.
  ISBN-13: 978-0-7506-8218-3 (pbk. : alk. paper)
  ISBN-10: 0-7506-8218-3 (pbk. : alk. paper) 1. Embedded computer
systems--Programming. 2. Debugging in computer science. I. Title.
  QA76.6.S572655 2007
  005.1--dc22
                              2007000130

British Library Cataloguing-in-Publication Data
A catalogue record for this book is available from the British Library.

ISBN-13: 978-0-7506-8218-3
ISBN-10: 0-7506-8218-3

For information on all Newnes publications visit our Web site at www.books.elsevier.com

07 08 09 10 11 5 4 3 2 1

Printed in the United States of America

To Thomas F. Mann
My dad, mentor and friend, and my original first reader.
For all my life he has encouraged me, counseling:
"To thine own self be true."

Table of Contents

Foreword

A few words from Jack Ganssle . . .

The oldest known book about engineering is the 2000-year-old work *De Architectura* by Marcus Vitruvius Pollio. One historian said of Vitruvius and his book, "He writes in atrocious Latin, but he knows his business." Another commented, "He has all the marks of one unused to composition, to whom writing is a painful task."

Does that sound like the last ten technical books you've read?

Engineers are famous for being very bright but also for lacking basic writing skills. Yet writing is still our primary means of communication, so we buy heavy tomes created without the benefit of basic grammar and often bereft of a coherent structure. Storyline? Character development? Forget it.

Welcome to a very different kind of technical book. Lisa Simone's work isn't the usual dreary tome stuffed with arcane wisdom buried beneath paragraph-length sentences seemingly written by someone just learning English as a second language. This is certainly the first embedded book with characters. The first with action, and with interesting and cool stories.

Bad code that makes a phone burst into flames?

What fun!

This is a James Patterson-style fast-paced book with dialog as close to gripping as one can imagine for a computer book. Its uniquely embedded focus twists together elements of hardware and software just as we engineers do in our daily design activities. One can't be understood without the other. Code makes the hardware smoke. That's unheard of anywhere but in the embedded industry.

Lisa weaves stories around deep technical issues. She's teaching the way humans have learned for 10,000 years. Most of us fought off sleep with varying levels of success in high school history classes. Who really wants to memorize the date of the First Defenestration of Prague or the name of Polk's vice president?

Yet now as adults we eagerly consume historical fiction (like James Michener) and real history assembled as a story (consider David McCullough). Cro-Magnon Grog taught his sons to avoid poisonous berries by telling them of uncles who died;

the Old Testament was passed down orally as a collection of stories rather than a dreary recitation of facts. Not properly casting an unsigned char sounds pretty dull, but when captured as a story - the interaction of people puzzling out a problem in the real-life settings we can all identify with - we're engaged and we learn the important lessons better.

This book is more real world than the standard text. Lisa shows how people are part of the solution and part of the problem. The concept draws on an oft-neglected axiom of the agile methods: people over process.

Despite the stories and character development, this is a textbook of a sort. There's homework. When Lisa asks you to stop and answer a question, do so! Think. Reflect. Surely Grog asked his sons questions to make them consider the lessons he'd imparted. We learn best by such interaction. Readers of Watts Humphrey's brilliant yet ineffably dull *A Discipline for Software Engineering* either do the homework and see their skills skyrocket; or read the book, skip the homework, and get no benefit at all.

Buried under the lessons, Lisa derives an important zeitgeist, a design pattern if you will, that should guide us in our work. It's one of creating readable work products: use cases, comments, requirement documents, and more. Though we need not emulate her use of story development in writing a report, we should and must abandon our traditional use of tortured English. Write interesting documents. Be lively and engaging. After 2000 years, it's time to leave Pollio's legacy behind and realize that if our readers are confused, frustrated or bored by what we produce, we're history.

Preface

A few years back, I was blindsided by a problem that was plaguing the creation of technology-based products and services. We don't teach our technical community of students, developers, computer scientists and engineers how to solve problems.

But I don't mean original technical problems, such as how to build a faster computer or how to diagnose cancer through advanced imaging analysis. Those "problems" are the types of technical and social challenges that got us into engineering and science in the first place.

What I mean is that there are unintended problems that plague us on our quest to create a portable artificial kidney or to deliver high-fidelity movies over a cell phone - problems that weren't supposed to be there in the first place.

I mean Bugs.

We're putting more bugs into our products without having the skills and expertise to take them back out again.

"Users are constantly uncovering defects in our . . . systems. We devote more time to fixing problems than developing new applications. We don't have time to do it right the first time, but we always have time to do it again." [1]

Companies are racing to deliver new products with cutting-edge features in less time - but quality has taken a back seat. Product complexity has skyrocketed. Software errors are estimated to cost the US economy $59.5 billion annually. [2] Software bugs in medical devices have killed patients.

You probably know that computers are embedded in everything these days, from cars and cell phones to washing machines and meat thermometers, but did you hear about the guy who had to reboot the exhaust fan on his stove?

The same bug that can cause that car's engine to stall, can also cause that cell phone to dial the wrong number, that washing machine to run too long, or that thermometer to prematurely indicate doneness.

What's the common element? The root cause?

When faced with a new symptom or bug, technical folks at every experience level are faced with the same question, "Okay, what do I try next?" How you answer that question is crucial. It is the difference between being an effective developer and debugger or a struggling team member. Wouldn't you like your next step to be a logical one instead of a random one?

It's all about your approach to problem solving.

Is This a Really Big Deal?

Maybe this wouldn't be such a *big deal* if we still had four-year product development cycles. A hint of this new reality emerged in the early '90s as a simple pressure to "compress the schedule a bit." But now we're shoehorning all that work into 18 months or less, and we've unfortunately done nothing to shrink the debugging cycle along with it.

Another *big deal* is that customers want neat new products, but they're also insisting on high quality. And now they're not afraid to demand last-minute changes, dictate terms, or drop you for another supplier seemingly on a whim.

But we must recognize that short schedules are permanent and just another constraint in an extremely complex dance that squeezes us all: the customers, the management, and the technical teams. How we address this challenge becomes the differentiator.

It's all about problem solving.

For several years I worked at major telecommunications companies as an embedded systems engineer and architect, designing mobile phones and leading mobile phone engineering teams. In the late 1990s, the telecom industry was in decline and 2000 was a terrible year; industry reports screamed of "carnage" and "tumbling stocks." It was "survival of the fittest" for companies large and small in the increasingly unforgiving economy driven by quarterly reports.

As it happens, we were late and the customer was screaming about poor product performance. I was on one of several daily status conference calls with engineers and managers from two continents and three times zones when reinforcements arrived. Five developers stood at my door ready for their first assignments. My management absolutely supported me by giving me these talented developers, but they were inexperienced in low-level embedded systems and new to this product. How was I going to quickly and effectively train them without letting the schedule slip any further? Should I drop everything to train them myself? Should I have the new engineers shadow the more experienced engineers? As an emergency problem solver myself, I knew the worse thing to do was to saddle my already stressed-out experts with novice engineers to mentor. But I might not have a choice.

Is This Problem Really Here to Stay?

Effectively designing and debugging embedded systems is solidly here to stay because nowadays computers are in everything. Only 2% of microprocessors end up in personal computers or servers; the rest are in devices such as your microwave, remote control, pacemaker, and even in (can you believe it?) some toasters.

Something I find amazing is that most folks don't know what an "embedded system" is, even though they own dozens of them. "Embedded systems pervade every aspect of life, often going unnoticed [by] the end user while at the same time enabling new activities, even a new quality of life" [3].

While finding and fixing software problems is challenging, debugging an embedded system is even worse. Everything is suspect - the hardware, the software, the mechanics, and the disposables if there are any (wait - batteries count). For better or for worse, we tend to design new systems from scratch because each is customized to a particular application. Washing machines don't need to perform floating-point math, and remote controls don't need high-resolution graphics displays; and there's generally not enough memory in the device for all that software anyway.

An advantage in embedded-systems development is that code-bloat is less prevalent than in the PC world. A disadvantage is that it makes code-reuse difficult, so fresh bugs are coded into algorithms that have been developed and verified a zillion times before for other devices.

Bugs. Lots more of them than ever before.

So How Can We Fix This?

What did I do with my new engineers? Without realizing it, I forced them to become more methodical. I couldn't focus on all the problems at once so I started asking tough questions about symptoms and behaviors. That kept folks busy until they got stuck again and came back to see me. Ping-ponging developers back and forth between my office and the lab made me feel a little guilty, but I didn't know the answers to most of their questions without investigating them myself.

But something wonderful happened. Many of the developers got smart - they started coming prepared with answers to my questions, even if the root cause of the bug still eluded them. I remember with great fondness a day when one of my engineers anticipated several steps in my thought process. In response to each of the successive questions I posed, he proudly whipped out one plot after another to prove he had identified, implemented, and validated a bug fix. Today, he is an experienced and respected embedded engineer at a major corporation, and I value the time I worked with him.

So what's the answer? Where do people learn to solve problems? No formal methodology is taught in schools. Formal on-the-job mentoring is an excellent source of training, but it is time-consuming and expensive.

So I wondered, could technical folks learn debugging skills the way I mentored my team? Each bug my engineers attacked was actually a mystery solved through methodical questioning. Something about that process had worked, because after a time, the ping-ponging between my office and the lab declined. The team became more adept at flushing out the bugs more quickly and we got a handle on the infestation.

As they say, the proof was in the pudding.

What Is In This Book

This book contains a series of technical mysteries for readers to solve. It's fiction, and it's nonfiction. "CSI for Embedded Systems" if you will. I've tried to make it as realistic as possible, from the way problems are found and reported through the pain of management pressure that "this must be fixed today." It also includes some traditional and nontraditional debugging methods and testing, and interactions with the folks from hardware, system test, and marketing.

But the technical stuff I'm trying to convey isn't as in-your-face as it is a consequence of the interactions of the Hudson Technologies engineers who solve each mystery. Oscar, Josie, Ravi, and Li Mei are technical folks who you've all met and worked with (or for). They're a typical engineering team. They have good days and bad; they have flashes of inspiration and at other times bang their heads in frustration on the bench after days of little sleep. While they're less dysfunctional than some teams can get, they feel the same pressure to nail the bugs amid sometimes frustrating team dynamics, broken or missing tools, some idiot customers and a stressful trade show demonstration. Despite it all, they learn and adapt, and blow off steam together after work.

Each mystery also begins with a *real-world bug* from Jack Ganssle's collection of embedded disasters. The disasters were selected to mirror the chapter mystery in symptom or in root cause. While some real-world bugs are humorous (temperature in a hot Texas town is reported at 505°F), others result in loss of life - a sobering tribute to the ease with which we randomly change 1s and 0s in our code.

Please note that this book will not teach you intricacies of a programming language, or attempt to trick you with pointers to functions returning arrays containing pointers to functions. It contains very little raw source code - only about 14 pages. Why? Because most powerful debugging techniques are language-agnostic. (Jumping straight into the software listings without a plan can derail the best intentions anyway.)

My goal is to tease you with a series of symptoms, clues, and realistic bugs, and provide logical steps to help you brainstorm root causes and propose targeted searches through the software and hardware.

How to Use This Book

This book can be enjoyed in a number of ways. It was written for folks of all experience levels, from students and new engineers to seasoned technologists. Each mystery is a window into the technical team's dynamics and it also provides a "Day in the Life" for folks curious about the field. Each mystery stands alone, although the Hudson team is the thread that weaves the stories together.

I invite you to grab a pencil and work through the examples to solve the mysteries before the team at Hudson Technologies does. Write in the margins. Think of other solutions. If you get stuck, work through the symptoms and the evidence side-by-side with Oscar's team and verify what they conclude. Or, just glide along with the team as they tackle one technical mystery after another. (I'll state upfront that none of the bugs are caused by minutiae like missing semicolons or misspelled variable names!)

While you get caught up in the mysteries, I hope some of the team's tricks get lodged in your brain and come floating back out when you find a similar bug or symptom in your own technical travels. When that happens, check out the appendix - everything you've learned is nicely summarized with references pointing back to each mystery.

It's all about your approach to problem solving.

Make your next step a logical one instead of a random one.

Contact Me

If you have comments or questions about these mysteries, or about embedded work in general, I'd be pleased to hear from you. Email me at *Lisa@simoneconcepts.com*.

For the latest info and updates, please visit *www.simoneconcepts.com/Phone_on_Fire*.

Lisa Simone

References

[1] Quality Assurance Institute. Reviews During Software Development. Accessed January 13, 2005 from *http://www.qaiusa.com/private/private_reviews.htm*.

[2] National Institute of Standards and Technology (2002). Software Errors Cost U.S. Economy $59.5 Billion Annually, NIST 2002-10, Accessed September 21, 2006 from *http://www.nist.gov/director/prog-ofc/report02-3.pdf*.

[3] Sami, M. What Are Embedded Systems? University of Lugano. Accessed January 13, 2005 from *http://www.alari.ch/EmbeddedSystems.htm*.

Acknowledgments

I am deeply indebted to friends and family who made this book easier and more enjoyable to create.

William J. Biessman reviewed the manuscript for technical details and posed many thought-provoking questions on the side effects of these mystery bugs during late-night email exchanges. When we first met, he showed up in my cube my first day on the job and without preamble began pouring his embedded-systems experience into my brain. On a different level, sometimes I appreciated more his uncanny knack for injecting the ideal comic relief needed to survive late-night debugging marathons.

Maria Schultheis suggested great ideas to improve the educational impact of Li Mei's List - collecting the lessons learned into a photocopy-able reference in the Appendix. She also provided constant encouragement, despite the insanity of my quest to write this book while everything else was going on.

Jack Ganssle has been an inspiration to me for more years than he knows. I got to know Jack through his fun-to-read books and newsletters on embedded systems. Who else provides the exact recommendations needed to keep *60 Minutes* from knocking at your door, or ponders how to explain a particular embedded sin to Saint Peter? I finally met Jack while pitching this book, and since then he has given me great advice and agreed to write the foreword. Jack also identified the excellent real-life bugs that introduce each mystery. Thank you, Jack!

I am grateful for the careful reading by several folks. Nicholas Chiarulli, Jr. diligently worked through the examples, wrote in the margins, and laughed in the right places. Hua Mao shared real-world experiences and big-picture feedback to make the learning a side effect of the fun. Tom Mann provided perceptive comments on almost every page of the draft manuscript, good-naturedly but seriously challenging me on everything from realistic team and management interactions to the use of the word "only" in the title. Linda Cheng and Nappinnai Sundarrajan shared insights from their own engineering experiences and career paths. Several folks made contributions (usually without realizing it) through their conversations, friendships, and sage advice: Joel Adler, Dean Bailey, Chuck Balas, Jack Gallagher,

Alice Higerd, Ziv Katalan, David Krause, Li Li, Annabelle Mann, Kathy Mann, Steve Petrucelli, Xiaorui Tang, and Mark Van de Walle. I must also thank the readers who provided positive feedback for me to pursue this book when a different version of one mystery first appeared in Embedded Systems Design magazine ("A Feynman approach to debugging," published by CMP, 2004). I am indebted to Terry Goodkind, who inspired the opening line to launch the adventure.

My thanks go to the folks at Elsevier and especially to my editor, Tiffany Gasbarrini, for her advice and enthusiasm for a different kind of embedded-systems book. She believed in the idea and championed it from the beginning.

Above all, to my wonderful husband, Al Simone, who has always supported my myriad adventures; this one was no exception. Finally, this book would not be possible without the constant inspiration and support of the following loved ones: Josie, Oscar, Kelly, Molly, BJ, WG, Sophie, Austin, Barney, and Hudson . . . for making anything and everything possible.

This is a work of fiction. The characters, products and situations depicted here are completely fictional. Any remaining (unintentional) bugs are purely my own. The figures in this text were created using Microsoft Visio and Excel; software was written in various flavors of C Language, assembly, and Parker 6K Motion Control Language.

"The shrewd guess, the fertile hypothesis, the courageous leap
to a tentative conclusion - these are the most valuable coin
of the thinker at work." - Jerome S. Bruner

"I haven't got much time to waste
it's time to make my way.
I'm not afraid of what I'll face
but I'm afraid to stay." - Madonna

About the Author

Lisa Simone wrote her first program on an Atari 800 in 1983, and cobbled together enough money *assembling* embedded systems to buy her first "modern day" computer (part by part) several years later. Since then, she has developed a wide range of products including medical diagnostic instruments, industrial automation and robotics, scientific measurement devices, and mobile phones and wireless sensor systems.

Lisa's experience includes pure research through product deployment. As a senior engineer at International Technidyne Corporation, she designed and developed hardware and software for portable blood coagulation devices. At Lucent Technologies and Motorola, she became a Distinguished Member of Technical Staff for her contributions to mobile phone architecture, integration and performance, and for managing and training new embedded engineers. She designed and implemented assembly-line automation for Spadix Technologies, and led bioinstrumentation research at Kessler Medical Rehabilitation Research and Education Corporation. She enjoys mentoring students and engineers, and volunteers as a judge and coordinator for student technology and engineering research paper and design competitions.

Today, Lisa is a biomedical engineering research professor at New Jersey Institute of Technology in Newark, where she teaches design, embedded, and biomedical engineering. She also designs devices and methods to understand human movement disorders caused by injury or disease, with funding from the National Institutes of Health.

Lisa received a B.S. in electrical engineering and an M.S. and Ph.D. in biomedical engineering from Rutgers University and the University of Medicine and Dentistry of New Jersey, and a Masters in the management of technology from the Wharton School and University of Pennsylvania.

Lisa currently resides in Bridgewater, New Jersey, with her husband and two high-maintenance cats. She loves scuba diving and underwater photography, reading and writing.

Chapter **1**

The Case of the Irate Customer: Debugging Other People's Code, Fast

*I*t was an odd-looking line of code, awkward in its form and syntax, dovetailed between well-formatted lines that marched up his computer screen. The pleasant left-and-right rhythm of indentation was marred by this single line, positioned brazenly flush with the left margin.

Not appropriate at all.

It was the offending line's placement that first caught his attention, as if it had been cut-and-pasted by mistake. Closer inspection added to his unease. The original author of this code was not the author of this line - a hack interloper had destroyed the beauty of this software. Oscar raked a hand through his hair as he pulled his focus away from the individual characters and syntax and let awareness of the code's function flood his brain.

It was a command to store a block of data into memory.

He scanned the comment section of the function and found no reference to the change. He wasn't surprised; someone writing sloppy code generally didn't pause to add comments.

But could this line be the source of the emergency, the reason why he'd been summoned back to work at 10 p.m. last night? And then spent the day alternately hunched over a lab computer and being dragged into various managers' offices to estimate when he could fix a bug that he hadn't yet had time to understand?

Three days before the final hardware and software were to be finished and delivered to manufacturing, the display on the Friend-Finder Communicator device suddenly turned red.

For no apparent reason.

Red.

He closed his eyes and rubbed them, thinking about symptoms of the bug and when it first manifested. The software was officially done and the device had been through System Test with no errors. Then the manufacturing test run was a resounding success and the devices had all passed inspection. The first production devices were packaged and shipped to the customer, who pulled one out of the box, turned it on and found the failure. A second exhibited the same failure. And then, one after another, each device powered up the wrong color.

Everyone had been stunned.

Normally the Communicators powered up an intense blue; a ring of LEDs cast a hazy glow around the edges of the display, making the words appear to float above the screen. With the lights out, animated graphics on the display seemed to rise out of a heavy mist.

Pretty cool.

But the customer's first Communicator had not powered up blue. Last night, Bright Lights' vice president had emailed a digital picture of five Communicators lined up on the table, all glowing an angry, accusing red.

The 7-million-dollar project was barely on schedule. Last-minute changes to the requirements had triggered a cascade of software problems and the lead developer on the project had suddenly resigned.

Oscar imagined the panic following that email. An initial management powwow, followed by realization that the lead programmer was no longer available, and then confirmation of the looming delivery deadline. Soon after, a frantic call to Oscar.

He picked up a Communicator and turned it on. The device happily powered up with the ring of blue LEDs around the display. What was going on?

Oscar turned his attention back to the software. The badly integrated and undocumented block-memory-copy code didn't appear to be related, but he launched the source-code version control program just to be sure, wondering when this software had been changed last. Each Change Request, or CR, to the file was tagged with the date, the reason for the change, the actual software change, and the version of official software that first included the change. He isolated the sloppy change to a software build over five weeks ago.

Since that change had been integrated, at least one premanufacturing run had been completed successfully. All the devices glowed blue. He made up his mind that the change was unrelated, closed the file and decided to look elsewhere.

This latest problem didn't make sense. The display ring could be configured to glow in several different colors, but those software routines were relatively old and unchanged. Oscar quickly typed the file name and the source code appeared; black characters on a white background. The font size was microscopic, a preference that allowed him to squeeze several pages of code on the screen at once.

The menu to select a display-ring color was pretty straightforward. He picked another of the bad Communicators, still glowing red, from the bench and selected the menu, and then scrolled to "LED Colors" and pressed the SELECT button. He was presented with a list of colors: BLUE, RED, YELLOW, GREEN, and WHITE. Blue was already highlighted at the top of the list. After he pressed SELECT, the ring of LEDs changed from red to blue and the device was happy.

Oscar was not.

Although he had done it several times already, he selected the "LED Colors" menu again and randomly changed the LED colors from blue to red, and then to green and white before changing the color back to blue. All of the menu options worked correctly. Just like they did every time System Test performed the automated test scripts for this product. Even when he power-cycled the device after changing the color, the bug did not reassert itself. Once the color was changed from the original offending red, the bug seemed to be gone for good.

He was losing his patience. As he stood up to get a can of soda, he saw Randy walk into the lab. Oscar sighed to himself and stood in place, wondering how many more management interruptions he'd have to deal with before he could get back to work. If only his cube had a door, he'd close it against the distractions and try to immerse himself in the problem.

This afternoon's meeting with the customer had not gone well and Randy still looked tense. "So, Oscar. Any progress?"

"No. I can change the color of this thing a hundred times and it works correctly. When I turn the power off and on again, it always starts up with the blue LEDs glowing."

"I've found that as well. Is it storing the new settings correctly? Is there any other way to change the color?"

Oscar kept his face expressionless to discourage conversation. His boss didn't write software and his attempts to provide debugging advice left Oscar feeling as if

Randy didn't have faith in him. "I haven't seen any other method yet." He paused to think about how the ring color data was stored. "I can use the debugger to set a breakpoint on a memory access of the color variable and see what function called it. That would reveal the call structure and nesting."

Oscar turned back to his computer and started to type, hoping that Randy would lose interest.

"Oscar, we need to talk about what this thing could cost us. Shipping this software late isn't going to be free. After you left the meeting, we talked about the financials. Bright Lights Corporation wrote penalty clauses into the contract, and they start kicking in if we don't get the manufacturing line on board for the product launch. And, by the way, Kenneth Anders didn't appreciate your feedback about his product."

"Penalty clauses? You signed a contract with penalty clauses in it and didn't say anything to me?" He tightened his grip on the Communicator and envisioned it exploding against the floor. He'd never met this Anders character before today, and he had immediately disliked him.

"I didn't sign this contract, Oscar. I got this over-budget project tossed on me just like you think it was tossed on you. You said the device wasn't that complicated. What's going on that you can't fix this? Can you have some of the people on your team work on this?"

Oscar felt his face getting warm. He didn't like his technical abilities questioned.

He knew he was tired and he pursed his lips to keep from saying something he would regret.

"How much is the penalty?"

"$10,000. Per day."

Oscar felt his stomach drop. "Per day? That's absurd!"

"I agree, but right now I'd rather not be negotiating the clause. Just avoiding it." Randy pulled out a lab stool and sat, staring at Oscar. After a moment, his voice softened slightly as he continued, "No one is going to be taking the money out of your paycheck, but my budget will feel the impact if we get hit with this, so you need to think about a different way to handle this. As you are fond of making clear to me, I am not an embedded developer."

Oscar's fingers froze over the keys. Without looking up, he said, "Well, Josie is between assignments and I asked her to work with the new employee tomorrow. She deserves a little down time to get caught up with email and whatever." He gestured offhandedly at the piles of trade journals stacked against the wall. "Maybe she'll have some time to brainstorm when she gets in tomorrow."

"Good. She seems to be making her mark. You've got a good engineer in her; don't screw that up." Randy stood. "Get this thing working."

Hours later, Oscar pushed back from the bench and expelled the breath he hadn't realized he'd been holding. It was 10:30 p.m. and he was alone in the lab, which was silent save for the periodic hum and pause of test equipment running automated scripts in the background. Closing his eyes, he pinched the skin at the bridge of his nose for a moment and then adjusted his glasses.

Red.

Perhaps foolishly, he'd asked in the meeting if *red* was really such a bad thing. After all, the requirements specified that the display should glow in a handful of colors. Did it truly matter if the device powered up red, and then the user changed it to blue using the menu?

Anders had crossed his long arms, protruding from monogrammed sleeves, and looked down his nose. *Clearly* Oscar did not understand.

"Throughout the movie," Anders informed him, "the color *blue* is synonymous with victory and camaraderie. The movie will end with the group of children defeating the alien using their Friend-Finder Communicators in the blue mode, signifying that they have met their challenges and successfully joined their power to defeat the invasion. *Red* means something else entirely, and not anything appropriate for the initial presentation of the Communicator to the potential audience." Then Anders had opened a folder and removed a thick document, ticking off the contract commitment dates for final software delivery and the start of manufacturing.

Oscar could still feel the narrow hard gaze Anders had directed at him. "The Friend-Finder Communicator *will* be available in stores on the day the movie is released nationally, and when it is powered up for the first time out of the box, it will glow blue. You may think this just a toy, but this is a multimillion-dollar product launch with a very hard deadline. So I suggest you go back and identify which of your colorblind employees has corrupted your software and is placing your job in jeopardy."

Since the meeting, exactly whose job should be in jeopardy was a question that periodically stole into his thoughts. His anger flared at the thought that they could be charged a penalty because the display was the wrong color. Picking up one of several Friend-Finders strewn across the lab bench, he pressed the power button. The display immediately glowed blue and the Friend-Finder logo scrolled across the display.

This one was working correctly.

He had no idea why.

After the heat of the meeting had dissipated, Oscar had to admit that it was actually a neat little gadget - sort of an organizer for young kids who used the Communicator to locate their friends when they wanted to play. The Communicator he was holding showed no other "kids" in his "neighborhood" although it constantly searched for other active Communicators within its area. The Communicators exchanged information about who was available, and the names of those friends were displayed. If Eduardo's Communicator was on, he could send Eduardo a smiley-face. Or, Oscar thought dejectedly, an invitation to go catch a much-needed beer at Molly's Pub.

He could also invite Eduardo to participate in a Challenge: some Hollywood thing where kids used the Communicators to save the earth. The Challenges were games that friends could play when several Communicators came into range. He recalled groups of engineers running through the halls, trying to form a perfect circle around a rogue Communicator. The rogue Communicators flashed and vibrated, spitting out confusing messages to the other devices. When a Challenge was successfully completed, all of the Communicators glowed blue and displayed a victory graphic from the upcoming movie. Cheers had echoed in the halls as the System Test group completed each challenge to validate the software.

So blue was actually a special color of sorts.

Despite his reluctance to admit it to that twit Anders, Oscar was more anxious to solve the mystery of the red display for the technical challenge, rather than for a fear that he would lose his job.

He would nail this bug.

Hudson Technologies was not going to pay penalties on his watch.

Josie grabbed her bag and swung the door of her car closed, locking it behind her as she headed through the parking lot to the employee entrance. The trees the company planted last year were blooming and the sun was shining. Another cold northeast winter was finally over and the "Garden" in Garden State had emerged from hibernation. It was an apt analogy for her own mental state. She'd finished a major project yesterday, and at Oscar's prodding had taken the rest of the day off to "see a movie or something." An appealing thought, but she'd spent the day lounging in front of a stack of DVDs eating sushi and popcorn, and then sleeping like a log through the night with one cat at her head and the other tucked behind her knees.

She felt significantly more clearheaded this morning than she had the last several months. She began ticking through the to-do's when she realized that she'd nearly forgotten that today was an auspicious day. Today was the new engineer's first day of work, and Josie was to be the Welcome Buddy for the week, making sure Li Mei was introduced around, given a tour of the building, and hosted for lunch the first day. That was more appealing than cleaning out the 521 emails in her inbox.

She smiled, thinking about her first nervous day at Hudson Technologies. She'd accidentally parked in the visitor's lot by the front of the building. Since they locked the front door at 5 p.m., she'd had to walk in the dark all the way around to the front from the employees' lot. Better make sure Li Mei doesn't make the same mistake. Deep in her thoughts, she swiped her ID badge to unlock the door and started up the long hallway to the cafeteria for the first cup of coffee.

Why am *I* nervous, she mused. This isn't *my* first day of work.

With coffee in hand, she made her way through the halls to her cubicle and stashed her bag under the desk, then wiggled the mouse to wake up her computer. She had long fallen into the habit of getting to work early to take advantage of the best debugging time of the day: the quiet before 9 a.m. She began scanning subject lines and deleting old mail and junk. "Daily bug report . . . Friday Lunch-and-Learn cancelled . . . Embedded Systems Newsletter . . . Fake Rolexes!" Ack! Now all she needed were some cheap pharmaceuticals and the morning email would be complete.

She took a long drink from her cup, and saw amid the electronic clutter an email from Oscar time-stamped late last night and set to high priority. Not good. It was addressed to the entire development staff, and the subject line said it all: "CODING STANDARDS ARE NOT A SUGGESTION." Ouch. Oscar was on the warpath again. Instinctively, she scrolled through her latest software in her head, checking for violations and hoping Oscar wasn't venting about something she had just submitted. The email continued with a terse, controlled rant about #defines, comments, and tab spacing, and concluded with an attachment containing the corporate coding standards document.

Probably best to steer clear of Oscar today.

As people began to arrive, Hudson Technologies came alive with typical development-environment chatter and shortly the front desk called. In the front lobby, she found a petite Asian woman who turned to her and smiled. Li Mei wore slacks and a button-down shirt and carried a red backpack. Josie felt herself smile back and introduced herself.

"Good morning, I'm Josie O'Neil. You're Li Mei? Did you find the place okay?"

"Yes, I am happy to meet you. I am Li Mei Cheng. The directions were fine. I arrived too early so I found a place to buy tea and waited for Hudson Technologies to open."

"Good. There are a lot of good places around for food and drink." She gestured for Li Mei to follow her out of the lobby and began pointing out various areas of the company as they walked. "Nearly half of the building is open cubical area. In the middle of the cubicle farm are the labs; that's where we do most of the development, although you can do a lot of software design, coding and documentation in your cube. We'll get your ID card programmed for our lab."

Josie introduced Li Mei around and they dropped off her backpack in her cube before continuing through the halls. Although the one-story building was not large, it took nearly half an hour to make the rounds from cubical to cafeteria, copier room to vending machines. Most importantly, Josie introduced her to the team's administrative assistant, Kathy. "This lady is a goddess," Josie gushed, making room for Li Mei to shake Kathy's hand. "Anything you need, she can do it or get it for you." After invoking a promise from Li Mei to stop by later, Kathy wished her good luck, and they continued on their tour.

As they chatted about Josie's projects, Li Mei asked her, "What work will I be doing? I have my degree in electrical engineering, but I like embedded software better." Then she quickly amended, "Although I am happy to work on anything."

Josie laughed. "Don't worry; you'll be doing different types of projects with a mix of hardware and software. We do a lot of contract work, but we also have a few of our own products that we design and sell, and some spec projects as well." She swiped her ID card and opened a lab door for Li Mei to follow her inside.

"I think Oscar is in our lab. He's a little swamped with a project, but I want him to know you're here. He'll probably meet with you today or tomorrow."

They found Oscar hunched over his bench with his chin cupped in one hand, using the other to step through code with a debugger.

"Hi, Oscar. Li Mei is here and I am giving her the grand tour."

Oscar startled and looked back, and then stood up and brushed the crumpled front of his shirt. Josie thought he looked exhausted.

"Hi, Li Mei. It's good to have you join our team." He shook her hand and then gestured around the room. "This is the embedded software lab; you probably remember from your interview. You'll have your own bench here with a computer, debugger and emulator . . . " His voice trailed off as he gestured haphazardly around the room. "Josie will help you get set up."

"Yes, I remember this lab." Li Mei looked around. The L-shaped room had high ceilings and several rows of benches that continued around the corner.

Oscar seemed at a loss for words, as if he were ready to say something else but couldn't make up his mind. Josie thought maybe it was just a bad time, and started to leave when he spoke quietly to her.

"Have you heard any new rumors about the Communicator project?"

She shook her head. "You sent me home early yesterday. What happened? Is this about the LEDs?"

He dropped his head and grunted. "We had a meeting with the customer yesterday. Idiot named Anders came in waving the contract and ranting about the colors. There's a penalty clause of $10,000 a day for late delivery so this thing is available for sale on the day the movie comes out.

"I know you need some down time, but I'd like to brainstorm with you this morning." He glanced at Li Mei for a moment, and continued, "Li Mei's first assignment is going to be the Meter Magic software that Benjamin didn't finish. I found another of his coding bombs in the software last night."

Josie flushed with relief; Oscar's email *hadn't* been directed at her.

Oscar turned to Li Mei and spoke with a little more determination. "Li Mei, I need to borrow Josie for a while. The Meter Magic project has some documentation that should give you an overview. How about Josie giving you everything and you start going through it?" Li Mei nodded. "You'll own all of that software, so you may as well get started on that."

Glancing back to the Communicator in his hands and fiddling with the buttons, he added, "And don't catch any bad habits from the code you're inheriting. I don't tolerate bad coding habits well."

Li Mei looked between them and stood up straighter. "I can do that. I will review everything and learn it and try not to bother you." She looked suddenly serious.

"Don't worry, Li Mei, we don't lock new employees in their cubes the first week," Oscar said. "You get a little ramp-up time before the torture begins." He gave her a tired smile and turned to Josie. "Would you get her set up?"

"Yes, let me do that now." She grabbed a notebook off a nearby bench and motioned to Li Mei. "Ready, Li Mei?"

"Okay." She stood quietly. "It will be good to start working on something today."

Li Mei's quiet response made Josie pause until she reminded herself how scary the first day at work can be, especially when your new boss looks rumpled and strung out. She promised herself to help Li Mei feel that she was a welcomed member of the team.

Josie returned to the lab a little later and relayed to Oscar that she had gotten Li Mei set up and occupied. Oscar pulled up a stool for her and gave her an overview of the software design. He hoped that her nods meant she understood his quick explanations. She looks so wide awake, he thought. He dreaded the idea of another long night in the lab if he couldn't get this resolved quickly.

Josie must have sensed his stress. "What is the timeline for this thing?"

He sighed. "That's the problem. The manufacturing line starts running the final software on Friday morning. That's a hard deadline. Working backwards, System Test needs 15 hours to run the test suite, so that must start first thing Thursday morning. The final software build with documentation takes three hours. No one is available to start a build at 4 a.m., but Mahesh will start one as late as 9 p.m. tonight and will stay overnight to baby-sit it. That gives us about 10 hours."

"Tonight?" Josie pulled back and stared at him. "The deadline is tonight and you let me go home early yesterday? I would have stayed."

"You needed a break, and it's probably better that you had a night to de-stress. This is my mess and I didn't want to drag anyone else into it." He didn't add that he thought he'd have it resolved by now.

"Well, what can I do?" She crossed her legs and rested her elbows on the bench. "What have you checked so far?"

He gathered his papers and tried to put his mind back into debugging mode. It felt like fuzz, but he forced himself to think back to the beginning. Staring at a point over her shoulder, without really focusing, he began. "First, I tried to verify the bug. I got one of the units straight from manufacturing that hadn't been turned on yet. When I powered it up, it glowed red. Bad." He smacked the bench with his palm to emphasize his frustration.

"I scrolled through the menus, and the color BLUE was highlighted. When I selected that, the LEDs turned blue. After I repowered the unit, the LEDs stayed blue. It only fails on first power up. I can't get it to fail otherwise." Josie nodded, and he took it as a sign to continue.

"Next, I checked all the latest changes to the software. The only Change Requests involved the graphics and menu options. All straightforward and none related to LED color." He explained the software changes in detail, but all had been code reviewed and tested before they were submitted.

Josie asked, "Do all the new units have this problem? Could they be using a bad programming unit at the factory to load the software?"

"All of the units glow red the first time. A bad device programmer; that's a good thought." He turned to the keyboard and tapped out a quick email. "They halted the manufacturing line, but I'll ask them to try another reprogrammer."

"What about the LED variable? The LED color must be stored in a variable somewhere, I imagine. Did you check accesses to the variable? Is it being over-written? Memory corruption?"

"All good thoughts. I checked the call-tree display already to see what software changes that variable. There are only three places. One is the menu system where the user can manually change the color using the menu option, one is a system test command that automatically changes the color using one of their test scripts, and the third is part of a routine that initializes the entire memory map. The initialization method isn't related to this, but I checked that all three places access the memory map correctly."

He went on to explain that each wrote the value to nonvolatile memory so changes would be saved after the device was turned off. FLASH memory on these portable devices was similar to storing data long term on a computer hard drive. "And before you ask," he added, "the variable is an unsigned char and every reference to it in the code is using an unsigned char, so there is no truncation or data loss."

Oscar wasn't sure if he was overwhelming her with information, but he continued anyway, plugging one end of a cable into the debugging port on the bottom of the Communicator, and the other to a port on his computer. "What I was about to do next was run the debugger with the Communicator connected, and explicitly check the LED color variable as the program runs to see if it's overwritten accidentally. I would have gotten to this sooner if I wasn't interrupted so many times by freaked-out managers. We might find that another array is accessed with a pointer out of bounds and that corrupts our variable in memory."

Josie queried him, "Pointer out of bounds - that's when you declare an array that's 10 elements long and then you accidentally write into the eleventh element?" She printed "array [10]" on her notepad. Oscar leaned over to look at her notes and nodded.

"If the array starts from zero, yes, that's a common bug. C language lets you get away with that very easily. I want to check for things like that." With Josie leaning in to see his monitor, he scrolled to the breakpoint menu. "If this is memory corruption, then the best way to detect it is using a watch point or breakpoint on the memory location itself."

"I use source breakpoints all the time."

Oscar wasn't sure she understood, so he explained as he typed. "This is a little different from a source breakpoint. When you set a source breakpoint, the program runs to that line of source code and then completely stops. Like using an axe." He looked at her. "And then you probably look at the value of that variable and then execute the software one line at a time while checking to see if that variable changes, right?"

Josie nodded.

"And then after a while, when you don't find anything, you let the program run again, and eventually it fails and you don't know why." As he expected, Josie was silent, but she seemed interested as she followed his keystrokes. "Using a watch point is less crude, although it will slow down the execution of the entire program as it runs." In the debugger, he selected "Add breakpoint" and configured it as a watch point on the variable **LED_Color**.

"The debugger will check after every line executed to see if this memory address has been accessed. That takes time, but it's thorough and the program doesn't stop running." He started the debugger, and then picked up the Communicator and watched it complete its initialization, or power-on-reset software.

When it finished, he scrolled through the menus to change the LED color and highlighted WHITE. As soon as he pressed SELECT, the debugger stopped and the breakpoint window was displayed. The **LED_Color** variable had been changed.

Oscar smiled and raised a finger. "Observe. This is proper behavior. We just verified that the debugger will stop when a new value is written to the LED color variable. Now that we know the debugger is set up correctly, we will try to get that watch point to fire when we're *not* trying to change the color. That way, if the debugger halts, we know something went wrong. Then we can look back in the debugger trace history to find out what part of the software was running when the improper memory access occurred." He reset the program and the debugger. The Communicator restarted and he randomly scrolled through the menu options. As Josie watched, he began changing every user-configurable setting in the device except LED color. The debugger continued to run.

Soon Josie picked up another Communicator and dragged the stool to her work bench, next to his. From the corner of his eye, he watched with satisfaction as she searched through the debugger options and found the watch-point settings to repeat what he was doing. After twenty minutes, she looked over. "It isn't tripping. That means that none of the user controls is causing this bug, right?"

"So it appears. And as I suspected. Otherwise, we would have found the bug before this. But it was important that we checked this for completeness."

She set the Communicator back on the bench. "Okay, what next?"

"Time to interview System Test. They were the first to see the bug." He checked the time on his cell phone. "The Late Risers should be in their lab by now."

The System Test lab was down the hall, but Oscar detoured to the vending machines for a soda.

"I don't know how you can drink soda in the morning." Josie grimaced, then intoned, "Coffee in the morning, no soda until after lunch."

"A little addiction you got there, eh?" Oscar said, unscrewing the bottle. Either coffee or soda tended to sit, forgotten, on the bench as he worked. He detested cold coffee, but warm soda wasn't that bad.

After swiping into the System Test lab, they found Eduardo fiddling with a piece of test equipment.

"Hey, Eddie, we have some questions about the Communicator. You made it go red first. What were you doing right before that?"

Eduardo turned. "I should be on the news; you're the third one to come and ask me that in the last two days. You're the big Technical Manager now. I give you interview number three, you give me the afternoon off?"

"Yeah, right, nice try. For all I know, you seeded that bug." Oscar leaned against the bench and rapped a Communicator against it. "We're not having a lot of luck, so we are trying to reconstruct what happened right when the bug first hit."

"Well, we did the same tests we've been doing all along. Everything is automated. The test scripts include loading the program into a Communicator, and then sending commands through the Communicator port to initialize and test the memory, like the LED color. More tests verify that the units can transmit and receive signals from other units, and simulate the user pressing keys to make sure the menu options work correctly for storing and recalling the Friends database information. You know, the information stored about each friend - name, favorite sound, friend color. After the test completed, we repowered up the unit and it came up red instead of blue." Eduardo shrugged. "Did you change the software? Because we test the blue and it works A-OK during our testing."

"We don't see any software change with the last load that could cause this," Josie said. "Can you make it happen again?"

"After the testing, we found it always happens just once. But we didn't see it until now because we don't turn the devices back on when the testing finishes without failure. The tests pass."

Josie began to doodle on her notepad. "If the color works at the beginning of your test but not after power cycle, then something in the test scripts is causing this. Can we run some of the scripts and see if we can narrow down when it happens?"

Eduardo's face clouded and he shrugged. "We can try. But we're in the middle of an automated test for another product. When that's finished, I'll jump you next in line. When I get your hardware jig and test scripts loaded, I'll call you."

Oscar nodded. "Thanks, Eddie. Just remember that we're in a crunch."

When Josie returned from her welcome lunch with Li Mei, she told Oscar that Li Mei had the documentation and that Ravi offered to take her to security and get her badge programmed for the lab doors. She seemed satisfied that Li Mei was occupied for now.

Oscar looked at her squarely. "Eduardo's not so happy that you implied his test scripts are the problem."

Josie replied defensively, "Well, I didn't mean for him to take it personally. But it could help us narrow down how it happens."

"That's true," he admitted. "Just be aware he's a little paranoid that he will be blamed for this, and with good reason.

"But I had an idea." He changed gears. "If our code is fine and this is a System Test script thing, then one solution is to power on every device from manufacturing and just change the LED color to blue using the menus. Then each device should never have the error again. On the other hand," his enthusiasm quickly dissipated, "someone actually has to do it. Labor costs might be more than the penalty clause until we identify the real bug. I gave Randy a heads-up and he's running the numbers."

"Ack - I sure as heck don't want to be that person!" Josie winced. "You ought to make that Anders guy from Bright Lights do it."

They shared a laugh, and Josie took a drink from her post-lunch diet soda. "Will you show me the Communicator software that System Test uses for the test scripts?"

Oscar turned back to the computer and pulled up a file. "System Test uses the Communicator hardware port as a back door to the software so they can simulate user keypresses and some system behavior. The commands are processed one at a time and ultimately get funneled down to a function called **process_ commands()**. The Communicator code and System Test scripts both use the same function." He pointed to the code in Figure 1-1.

> **Reader Instructions:** Before continuing, review the function in Figure 1-1. Be forewarned - the code is not well written, but it works. Figure out the overall functionality first. Do you see anything suspicious? Do the #defines help or hurt your understanding? Don't give up - Oscar and Josie will step through the important issues.

"The original author could have used a few more comments and #defines for clarity, but it isn't too hard to understand if you look at the big picture. The function receives a command, **cmd**, and the switch uses it to control what code is executed. The function also takes a pointer to an array of characters that contains data to be read or stored, an offset, and a return code for status of the operation."

```
/* Friend-Finder Communicator Software Code Listing (Partial)           */
#define MAX_NUM_FRIENDS        10
#define MAX_FIRSTNAME_LEN      15
#define MAX_LASTNAME_LEN       15
#define SIZEOF_FRIEND          (MAX_FIRSTNAME_LEN + MAX_LASTNAME_LEN + 3)
#define MEM_START              0x00
#define FRIEND_AREA_START      0x00
#define PARAMETER_AREA_START   (FRIEND_AREA_START + MAX_NUM_FRIENDS*SIZEOF_FRIEND)
#define FIRSTNAME              0x00
#define LASTNAME               (FIRSTNAME + MAX_FIRSTNAME_LEN)
#define FRIEND_TYPE            (LASTNAME + MAX_LASTNAME_LEN)
#define FRIEND_COLOR           (FRIEND_TYPE + 1)
#define FRIEND_SOUND           (FRIEND_COLOR + 1)
#define DISPLAY_LEDS           PARAMETER_AREA_START
#define TRANSMIT_PARAMS        (DISPLAY_LEDS + 1)
#define BATT_CHARGING_PARAMS   (TRANSMIT_PARAMS + 10)
#define POWER_CONTROL_PARAMS   (BATT_CHARGING_PARAMS + 6)

enum Colors {LED_RED, LED_BLUE, LED_YELLOW, LED_GREEN, LED_WHITE};

/* Structure for Friend-Finder Database Entry */
struct friend_data_entry_type {
    char first_name[15];
    char last_name[15];
    char friend_type;
    char friend_color;
    char friend_sound;
    char friend_available;
} friend_data;

struct friend_data_entry_type Friends[MAX_NUM_FRIENDS];
char LED_color;
char RxTx;
char Battery;
char Power;

void process_commands(char cmd, int offset, char *data_ptr, char *return_code)
{
    char status;
    int  len;
    switch (cmd)
    {
        case 0x01 :  len = (strlen(data_ptr) <= (MAX_FIRSTNAME_LEN-1))
                          ? strlen(data_ptr) : (MAX_FIRSTNAME_LEN-1);
                     status = store_data (FRIEND_AREA_START
                         + (offset*SIZEOF_FRIEND) + FIRSTNAME, data_ptr, len+1);
                     break;
        case 0x02 :  len = (strlen(data_ptr) <= (MAX_LASTNAME_LEN-1))
                          ? strlen(data_ptr) : (MAX_LASTNAME_LEN-1);
                     status = store_data (FRIEND_AREA_START
                         + (offset*SIZEOF_FRIEND) + LASTNAME, data_ptr, len+1);
                     break;
        case 0x03 :  status = store_data (FRIEND_AREA_START + (offset*SIZEOF_FRIEND)
                         + FRIEND_TYPE, data_ptr, CHAR_SIZE);
                     break;
        case 0x04 :  status = store_data (FRIEND_AREA_START + (offset*SIZEOF_FRIEND)
                         + FRIEND_COLOR, data_ptr, CHAR_SIZE);
                     break;
        case 0x05 :  status = store_data (FRIEND_AREA_START + (offset*SIZEOF_FRIEND)
                         + FRIEND_SOUND, data_ptr, CHAR_SIZE);
                     break;
        case 0x06 :
store_data (offset,data_ptr, sizeof(friend_data));  break;
        case 0x10 :  status = store_data (DISPLAY_LEDS, data_ptr, CHAR_SIZE);
                     break;
        case 0x20 :  status = store_data (TRANSMIT_PARAMS, data_ptr, 10*CHAR_SIZE);
                     break;
        case 0x21 :  status = store_data (BATT_CHARGING_PARAMS, data_ptr, 6*CHAR_SIZE);
                     break;
        case 0x22 :  status = store_data (POWER_CONTROL_PARAMS, data_ptr, 3*CHAR_SIZE);
                     break;
        default :    break;
    }
    *return_code = ERROR;
    if (status == 1)      *return_code = OK;
}
```

Figure 1-1 Software Listing to Process Commands.

Oscar interrupted himself. "Oh, and the factory also returned my email - the programming devices are not the problem. Apparently happens on every one of them."

"Too bad," Josie sighed.

"Anyway, each command stores some data. They look confusing but they're similar. The first several compute the length of the incoming data string first, and then use the conditional operator '? :' to truncate the length to the max allowable length if it's too long for that database entry. That's shorthand for an if-then-else statement."

She nodded her understanding and he pointed to the first several. "Look at the #defines. These write entries into the Friends database, stuff like first and last name. Each command is a different database entry or parameter. At the bottom of the switch are commands for parameters like battery operation and transmit and receive power. **LED_color** is the 8-bit variable that we're interested in."

He straightened on the stool and turned to her. "Notice anything interesting, without knowing more about the code?"

Josie peered at the screen. "Yes, **status** isn't initialized. If **cmd** isn't handled in the switch statement, the return code argument in **process_commands** will be indeterminate."

"Good catch. I didn't see that. But that's not what I was looking for."

She continued to inspect the function. "Command 0x06 is what prompted your hate email last night, right?"

Oscar grunted. "That would be the cause." He'd forgotten about the email he had dashed off in frustration early this morning. "Anything else?"

"The command after that is the one that writes the LED color. Coincidence?"

"It could be. But remember that those two chunks of code don't run consecutively. Only one command is executed each time through the function. The switch statement controls the execution based on the input command. But you're on the right track. Check out the #defines for the memory map. Don't get caught up in the details."

She scrolled to the top of the file and located the #define for **DISPLAY_LEDS**. "All of the parameters are located in memory right after the Friends database. So if anything overwrote the end of the database, it would overwrite the LED color."

"Bingo."

It was later in the afternoon by the time Eduardo finally found them. Oscar and Josie had walked through the code several times and he was reasonably certain that they both understood the calling structure. When they arrived back in System Test, Eduardo had the Communicator in the test jig and he started the automated test scripts. They watched the display scroll through menus and access data, seemingly controlled by invisible hands. Oscar's thumbs felt tired just looking at the rapidly changing screen. Shortly after testing started, the LEDs glowed blue, followed by red, green, yellow, white, and finally back to blue. After several minutes, the test completed and the device powered down. Eduardo removed it from the test jig and handed it to Oscar.

"You can do the honors."

Oscar pressed the power button.

The display graphic flickered and the LEDs glowed red. He groaned, but quickly reversed himself. "This is actually a good thing - if it's red now, then we can work backwards. Josie and I took a look at the software used to write the memory, and the Friends database is located right before the LED color variable. We'd like to run the test scripts one at a time, and stop just past the point where the LEDs are tested but before the database is written."

Eduardo gestured for the Communicator and placed it back in the jig. "Okay, honored debugging guru, we run to the database. You think the database is the problem?"

Oscar ignored the wisecrack. "I do. We'll check the variable right before the database is written, and then check it again afterwards. I suspect it will be changed."

Eduardo halted the scripts after the LED test was completed and checked the **LED_color** variable. Its value was 1; the LEDs were configured for blue. He restarted testing, and then rechecked the variable and turned to Oscar.

"The variable has changed to zero, but the LEDs are still glowing blue. What does this mean? Is it the bug?"

Oscar felt a smile appear on his face and the stress seep from his shoulders. That singular point when he knew he had a grasp on the tail of the bug was always an intense relief.

The value for red LEDs was 0.

"It surely is the smoking gun. And, no offense, but I am extremely pleased that your script pulled the trigger." Oscar saw Eduardo tense defensively, and he gestured for patience.

"Eddie, I still don't know where this bug is, but your test script flushed it out. This is a major victory, but we still haven't solved the problem. Don't get bent, okay?"

"System Test gets blamed for a lot, Oscar. Don't let my manager know about this until you know for sure where the problem is. I don't need more frustration."

"You're great, Eddie. Josie and I are going to go check the software that's used to store the database elements, and we'll get back to you."

"Okay, I got the System Test script manual for the database that describes the test. The script fills the database completely. It doesn't hold much - only ten friends' worth of data." Oscar showed her the list. "And the friend color is different from LED color. Each kid can pick a color for his name."

Each Friend Finger Database Entry:

First name	15 bytes
Last name	15 bytes
Friend Type	1 byte
Friend Color	1 byte
Friend Sound	1 byte

Josie spread the document on the bench and drew a diagram (Figure 1-2). "Let me understand - all commands get funneled down to **process_commands()** to change the LED color during normal operation, right?" He nodded, noting she'd also identified that the color could be changed by initialization. "So all the System Test script commands that come in through the test port ultimately end up using the same functions as the internal menu system."

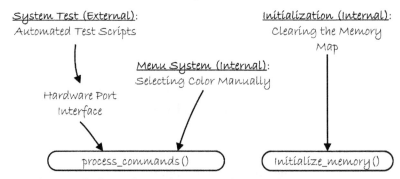

Figure 1-2 Three Methods to Change the LED Color Variable.

"Yes. The Communicator software that writes the LED variable shouldn't know the difference. We're all using code like this - it takes five commands to fill one database entry. I built a debugging function to fill the entire database, but this software doesn't cause the problem."

```
construct_cmmd(0x01, 0,"Kid1Firstname");
construct_cmmd(0x02, 0,"Kid1Lastname");
construct_cmmd(0x03, 0,"0");
construct_cmmd(0x04, 0,"0");
construct_cmmd(0x05, 0,"0");

construct_cmmd(0x01, 1,"Kid2Firstname");
construct_cmmd(0x02, 1,"Kid2Lastname");
construct_cmmd(0x03, 1,"1");
construct_cmmd(0x04, 1,"1");
construct_cmmd(0x05, 1,"1");
```

Oscar pulled down the front of his shirt to straighten it, noting despondently that the effort was futile. His earlier confidence was beginning to fade as the deadline approached. The building was nearly cleared out for the night, but he noted with relief that Josie didn't fidget to pack up and leave.

If everyone used the same command, he pondered, why couldn't they duplicate the problem from the internal Communicator interface? They'd have to do it differently.

"That code looks okay. Is there any easy way to know exactly what parts of memory got changed?" She picked up the soda bottle and started to drink.

"Well, we can dead beef it."

Josie nearly choked on the soda. "'Dead beef it'? What the heck does *that* mean?" She wiped her mouth with the back of her hand.

"Ah, you young ones. There is much you do not know." He smiled sublimely and assumed the crossed-arms pose of the wise instructor. Josie rolled her eyes, preparing for the lecture.

"Dead beef is one of the interesting phrases that can be formed using only the available hexadecimal characters A through F. DEADBEEF. A trick to checking what parts of memory have been written is to first lay down a recognizable fill pattern in memory. DEADBEEF is a historical pattern. After the program in question runs, it's easy to see where the pattern is broken." He smiled down his nose at her and said in his best pompous authority voice, "We shall DEADBEEF the memory and retest the software."

Josie laughed as he opened a file and began to code. "Okay, so how do we do this 'dead beef' thing?"

She seemed genuinely interested in the vintage method and he decided to teach her more about detecting memory corruption.

He asked, "You read a lot of sci-fi, don't you?" Seeing her nod, he continued, "Who's your favorite science-fiction author?"

The question caught her off guard, but she was quick to reply, "Marion Zimmer Bradley. No wait, Piers Anthony is good too, and Isaac Asimov."

"Any others?" His eyes never left the screen as he typed.

"Well, how about Orson Scott Card?"

Oscar soon had another new debugging function. This one filled the entire memory with the repeating phrase DEADBEEF. He placed the call directly before his function to populate the database.

```
debug_init_mem_map_deadbeef();
debug_fill_friend_memory();
```

He commanded the debugger to run to the second function, and then showed Josie the memory filled with the repeating phrase (Figure 1-3). The memory viewer showed 20 columns and 24 rows of data in hex on the left, with the ASCII representation of each on the right. "See, the four ASCII characters for 'DEAD' are stored as 0x44, 0x45, 0x41, 0x44." She nodded for him to continue.

```
      0 1 2 3 4 5 6 7 8 9 0 1 2 3 4 5 6 7 8 9     01234567890123456789
 0:  4445414442454546444541444245454644454144     DEADBEEFDEADBEEFDEAD
 1:  4245454644454144424545464445414442454546     BEEFDEADBEEFDEADBEEF
 2:  4445414442454546444541444245454644454144     DEADBEEFDEADBEEFDEAD
 3:  4245454644454144424545464445414442454546     BEEFDEADBEEFDEADBEEF
 4:  4445414442454546444541444245454644454144     DEADBEEFDEADBEEFDEAD
 5:  4245454644454144424545464445414442454546     BEEFDEADBEEFDEADBEEF
 6:  4445414442454546444541444245454644454144     DEADBEEFDEADBEEFDEAD
 7:  4245454644454144424545464445414442454546     BEEFDEADBEEFDEADBEEF
 8:  4445414442454546444541444245454644454144     DEADBEEFDEADBEEFDEAD
 9:  4245454644454144424545464445414442454546     BEEFDEADBEEFDEADBEEF
10:  4445414442454546444541444245454644454144     DEADBEEFDEADBEEFDEAD
11:  4245454644454144424545464445414442454546     BEEFDEADBEEFDEADBEEF
12:  4445414442454546444541444245454644454144     DEADBEEFDEADBEEFDEAD
13:  4245454644454144424545464445414442454546     BEEFDEADBEEFDEADBEEF
14:  4445414442454546444541444245454644454144     DEADBEEFDEADBEEFDEAD
15:  4245454644454144424545464445414442454546     BEEFDEADBEEFDEADBEEF
16:  4445414442454546444541444245454644454144     DEADBEEFDEADBEEFDEAD
17:  4245454644454144424545464445414442454546     BEEFDEADBEEFDEADBEEF
18:  4445414442454546444541444245454644454144     DEADBEEFDEADBEEFDEAD
19:  4245454644454144424545464445414442454546     BEEFDEADBEEFDEADBEEF
20:  4445414442454546444541444245454644454144     DEADBEEFDEADBEEFDEAD
21:  4245454644454144424545464445414442454546     BEEFDEADBEEFDEADBEEF
22:  4445414442454546444541444245454644454144     DEADBEEFDEADBEEFDEAD
23:  4245454644454144424545464445414442454546     BEEFDEADBEEFDEADBEEF
24:  4445414442454546444541444245454644454144     DEADBEEFDEADBEEFDEAD
```

Figure 1-3 Hex and ASCII Dump of DEADBEEF on Patterned Memory.

After stepping over the **debug_fill_friend_memory()** function, the DEADBEEF pattern was overlaid with a hastily assembled Top 10 of Who's Who in Science Fiction and Fantasy (shown in Figure 1-4).

Josie laughed as she scanned the right side of the memory viewer on Oscar's monitor. "Pretty smooth! Oh, Dan Simmons - I forgot him!"

Each character of the friend's names wiped out the underlying DEADBEEF character, leaving snippets of the repeating phrase to show through among the

```
        0 1 2 3 4 5 6 7 8 9 0 1 2 3 4 5 6 7 8 9   0123456789012345678 9
 0: 49736161630045464445414442454541736969D6F   Isaac EFDEADBEEAsimo
 1: 76004546444541444245303030506965727230046   v EFDEADBE000Piers F
 2: 4445414442454546416E74686F6E4790044454144   DEADBEEFAnthony DEAD
 3: 424545313131526F62657274004541444245454 6   BEE111Robert EADBEEF
 4: 4453696C766572626576267004245454632323244   DSilverberg BEEF222D
 5: 6F75676C61730044424545464444541646461 6D7300   ouglas DBEEFDEAdams
 6: 44454144424545464433333335261790044454144   DEADBEEFD333Ray DEAD
 7: 4245454644454142726164627572790042454546   BEEFDEABradbury BEEF
 8: 4445343434506869 6C6970204B2E004644454144   DE444Philip K. FDEAD
 9: 4469636B0045414442454546444541353535 4765   Dick EADBEEFDEA555Ge
10: 6E6500444245454644454414442526F646465 6E62   ne DBEEFDEADBRoddenb
11: 6572727900454144363636 4D6172696F6E205A69   erry EAD666Marion Zi
12: 6D6D65720045427261646C657900454644454144   mmer EBradley EFDEAD
13: 4237373744616E0042454546444541444245453   B777Dan BEEFDEADBEES
14: 696D6D6F6E6730046444454144424538383834F7273   immons FDEADBE8880rs
15: 6F6E2053636F747400454543 61726400454546   on Scott EEFCard EEF
16: 444541444245453939[01]040505040305030200    DEADBEE999[ ]
17: 0001010202030306060645464445414442454546       [ ]EFDEADBEEF
18: 4445414442454546444541444245454644454144   DEADBEEFDEADBEEFDEAD
19: 4245454644454144424545464445414442454546   BEEFDEADBEEFDEADBEEF
20: 4445414442454546444541444245454644454144   DEADBEEFDEADBEEFDEAD
21: 4245454644454144424545464445414442454546   BEEFDEADBEEFDEADBEEF
22: 4445414442454546444541444245454644454144   DEADBEEFDEADBEEFDEAD
23: 4245454644454144424545464445414442454546   BEEFDEADBEEFDEADBEEF
24: 4445414442454546444541444245454644454144   DEADBEEFDEADBEEFDEAD
```

Figure 1-4 Hex and ASCII Dump of Good Code on Patterned Memory.

famous authors' names and the three friend's numbers following each entry. He showed her that each name was terminated with a NULL character, or zero, so the software could detect the end of each name string. When they reached the bottom of the list, his humor evaporated. Data from the last author was overlaid straight to the end of the Friends database memory area, but no further.

The **LED_color** variable remained unchanged.

Still 0x01.

Still blue.

Josie pointed to the printed squares on the right. "I don't understand. **LED_color** is a number but it's printed as a square, but numbers that are in the Friends database like Orson Scott Card's '999' are printed as real numbers."

"Remember, this is a debugging display - memory contents are not always in a format that is easily readable. Values less than Hex 20 are nonprintable characters, and that square you see is just a placeholder for a nonprintable character." He pointed to the corresponding data location on the left and circled the **LED_color** variable. "For these, you have to look at the actual hex value on the left side of the display. It is correct; the **LED_color** value is 0x01 and that means blue." The System Test script was not the source of the bug.

Josie turned to face Oscar. "Oh, I get it. You thought the software would overwrite the end and change that value to a zero."

"Yes. Back to Eduardo."

"Hey, Eduardo, thanks for staying late. We just duplicated what the System Test script sends to the Communicator to fill the Friends database. It doesn't corrupt memory like your script did, and it doesn't cause the LED color problem. Are you sure that's the test you are running?"

Eduardo tossed his hands in the air. "I know what tests are being run. I know this system inside and out! The only thing that changed with that test is that we requested an update to your software a couple months ago for a more efficient command to fill memory. That was tested and released. Last week we changed the contents of the Friends database to test another CR, but that has nothing to do with the code or the LED color."

Oscar stopped short.

"You changed something?"

"Not the software - just the data we load into memory."

"Show me." Oscar walked him to the computer and Eduardo pulled up the source code for the test script that successfully reproduced the failure (shown in Figure 1-5).

Like a well-shot arrow, Oscar's finger was already smudging the screen. "This constructs a test string to use command 0x06? We are not using that command. Is this test script calling the new Communicator code you mentioned?"

"Yeah, that's it. Before, we had to fill a friend's entry one field at a time and it took a lot of code, so we requested a command to write an entire entry with one line. Is that related to the problem?"

"I don't know, but we need to redo our debugging test using that command. From our end, it's poorly implemented." He didn't add that the system-test

```
/* System Test Automated Script (partial code listing)
   Product:   Friend Finder Communicator
   Test #12:  Fill database memory
   Function:  Send 10 commands with sample entries.              */

void fill_database(void)
{
    int  i = 0;

    construct_cmmd(0x06, 33*i++, "First          Last          111");
    construct_cmmd(0x06, 33*i++, "First          Last          222");
    construct_cmmd(0x06, 33*i++, "First          Last          333");
    construct_cmmd(0x06, 33*i++, "First          Last          444");
    construct_cmmd(0x06, 33*i++, "First          Last          555");
    construct_cmmd(0x06, 33*i++, "First          Last          666");
    construct_cmmd(0x06, 33*i++, "First          Last          777");
    construct_cmmd(0x06, 33*i++, "First          Last          888");
    construct_cmmd(0x06, 33*i++, "First          Last          999");
    construct_cmmd(0x06, 33*i++, "First          Last          111");
}
```

Figure 1-5 New System Test Software to Write Friends Database Information.

implementation with hard coded "33s" was about the same level of quality as the 0x06 command. Instead, he spun on his heels and marched out of the lab. Josie shrugged at Eduardo and followed Oscar out.

Oscar ranted as he headed back to the embedded lab. "Did you see that code? That's another thing that gets me. Just because software running in System Test isn't shipped, it doesn't mean it shouldn't be well tested and documented. And now they are using custom commands to test our memory and bypassing the very commands that we are using. That defeats the entire purpose of testing!" He stomped to his bench.

While he fumed, Josie slid up to her own bench and keyed up the file containing the new command. The 0x06 command was similar to the others in the function. Rather than receiving one Friends parameter at a time, it appeared to receive and write an entire entry at once. **Friends** was the global variable containing the entire database; 10 entries of type **friend_data_entry_type**.

At first glance, it appeared to be okay.

Oscar tried to calm down as he changed his debugging program to fill memory using command 0x06, mimicking the system-test scripts as closely as he could. He wanted fiercely for this test program to cause the bug. The deadline was upon them, and it would be sweet to nail the author of this hack. He glanced at Josie, who was still scrolling through the code. After his change compiled, he ran the program and opened the output file (shown in Figure 1-6).

```
      0 1 2 3 4 5 6 7 8 9 0 1 2 3 4 5 6 7 8 9    01234567890123456789
  0:  4973616163202020202020202020204173696D6F    Isaac           Asimo
  1:  762020202020273202020303030506965727320     v        000Piers
  2:  202020202020202020416E74686F6E792020202020             Anthony
  3:  202020313131526F626572742020202020202020      111Robert
  4:  2053696C766572626572672020202020203232324   Silverberg      222D
  5:  6F75676F61732020202020202020204164616D7320   ouglas          Adams
  6:  2020202020202020203333333352617920202020          333Ray
  7:  20202020202020204272616462757279202020202         Bradbury
  8:  2020343434345068696C6970204B2E202020202020      444Philip K.
  9:  4469636B20202020202020202020203535354765    Dick            555Ge
 10:  6E65202020202020202020202020526F646465E62   ne          Roddenb
 11:  65727279202020203636364D6172696F6E205A69    erry       666Marion Zi
 12:  6D6D65722020427261646C6579202020202020      mmer   Bradley
 13:  2037373744616E20202020202020202020202053     777Dan            S
 14:  696D6D6F6E732020202020202020203838384F7273   immons          8880rs
 15:  6F6E2053636F7474202020204361726C7264202020   on Scott   Card
 16:  2020202020202039393900040505040305030200          999
 17:  000101020203030606064546444541442454546                 EFDEADBEEF
 18:  4445414442454546444541444245454644454144    DEADBEEFDEADBEEFDEAD
 19:  4245454644454144424545464172696F6E454546    BEEFDEADBEEFDEADBEEF
 20:  4445414442454546444541444245454644454144    DEADBEEFDEADBEEFDEAD
 21:  4245454644454141424545464445414442454546    BEEFDEADBEEFDEADBEEF
 22:  4445414442454546444541444245454644454144    DEADBEEFDEADBEEFDEAD
 23:  4245454644454141424545464445414442454546    BEEFDEADBEEFDEADBEEF
 24:  4445414442454546444541444245454644454144    DEADBEEFDEADBEEFDEAD
```

Figure 1-6 Hex and ASCII Dump of System Test Code on Patterned Memory.

"Got it." He smacked the bench with his fist. A tiny smile crept onto his face. "Check it out - look at the difference in memory patterns for the good code and the bad code. What do you see?"

Josie leaned over to look at his display, where he had both memory patterns displayed side by side. He watched her take in the differences. Even though the System Test software completely overwrote the pattern in between the names, the first and last names still appeared at the same place in each memory map. That apparently didn't affect how the software ran.

She straightened. "Sheesh, I see it. After the '999' in the System Test case, there is a blank space before the DEADBEEF starts again. That's exactly where the **LED_color** variable is stored. Somehow that new command ends up stepping on memory one byte beyond where it is supposed to. That's the bug!" She furled her eyebrows. "But that doesn't explain why the LED didn't turn red when Eddie ran this piece of software for us."

"That's what we must discover." Now that he had nailed the bug down, it needed to be unraveled and fixed. Normally he enjoyed the process of understanding exactly how it worked and then implementing the fix, but they were out of time.

His computer beeped and launched an email notification. Just as they proved that they could invoke the bug, Mahesh was expecting the completed software change to launch the official build. How had it gotten to be 9 o'clock so quickly? They could probably delay Mahesh a short while, but they hadn't even proposed a fix or created a test build to verify it.

With a sigh, Oscar ignored the email and returned to the memory dump. As he pondered, Eduardo's unanswered question from that afternoon resurfaced. If the System Test scripts had changed the LED color variable but the LED still glowed blue, were they looking at the right variable? He glanced at the Communicator beside him; they had just changed the software using the System Test scripts and the LED glowed blue.

The symptom replayed in his brain. The LED color variable was changed to RED, but the LEDs still glowed blue.

Suddenly he remembered part of the original symptom. They had to power cycle before the LED color changed. He explained his train of thought to Josie as the Communicator rebooted. But instead of the red he anticipated, it again powered up blue.

Oscar was stunned. "I am getting a little crazy here - we need to stand back and be logical." He put the Communicator on the bench and raked his hands through his hair. "We have a couple of avenues to explore. First, why is command 0x06 overwriting the LED color variable? And second, why doesn't that reproduce the problem?

"Command 0x06 writes a string of data from System Test," he said, pointing to the **fill_database()** function from the System Test script code, "and the string is not too long. Look, the string ends with the three parameters."

Josie asked, "Since they are writing a string rather than a number, is the software automatically storing a string terminator at the end like the NULL character?"

"That's a thought, but look at the code for command 0x06. It doesn't use terminators; it explicitly sends the length of the data as the third argument. It sends the size of the **friend_data** structure." He started to think, and scrolled up to the declaration. "The Friend-Finder Database Entry structure is 15 plus 15 plus 4 bytes long. 34 bytes."

Suddenly Josie dipped her head backwards and exclaimed, "Oscar - I get it! I know what's happening! System Test is using the data structure to derive the memory size instead of using the #defines for size." She pointed to the #defines at the top of the file. "Each **SIZE_OF_FRIEND** entry is 33 bytes, not 34 bytes!"

He quickly scanned the file and found that she was right. Josie had seen the tiny clue that he had missed. "Check it out - the **friend_data** data structure is one byte larger - the last byte is a temporary byte that's only used when the program is running. It tracks if a friend is currently available. It's not a part of the permanent database in FLASH."

She continued in a rush. "The new command that System Test started using only supplies 33 bytes worth of data, like it's supposed to. But whoever coded command 0x06 used the **sizeof(friend_data)** to derive the size, and it accidentally told the software to write 34 bytes. Who knows what's in that last byte of data? It could be the mystery value for RED."

He stopped. None of this explained why a power cycle was needed to invoke the bug, and then his brain flooded with complete understanding. Suddenly everything fell into place and he pushed out his stool to pace with newfound energy.

"Okay, see if this makes sense. First, System Test requests the new 0x06 command and begins using it. For some reason, the extra garbage byte that gets written is a zero, just like we reproduced here. If the **LED_color** value is zero, the LED color is red, right?"

Josie was nodding. Everything had been working correctly, completely by accident. Her Communicator now lay forgotten on the bench as she followed his pacing.

"Next, Eddie said that they changed the contents of the Friends database. Even that shouldn't affect that final critical garbage byte, but who knows. They did something that changed RAM - whatever was stored in the variable directly after that command buffer." He continued walking back and forth between the rows of benches, trying to remember the rest of the System Test program, but then decided it didn't really matter. "And that single byte became mystery data byte number 34."

He stopped pacing and faced Josie, his face hot. "And the reason that we don't see the LED change color immediately is that these commands are writing the data to FLASH memory. They commanded a change to the database, so the database was updated in FLASH, and then the local database variables were updated. The command accidentally changed the LED color variable in FLASH, but not local memory. The critical variable, **LED_color**, didn't get changed immediately. The only time we saw the change is after a power cycle."

"And what happens after a power cycle?" He looked at her, expecting to see her sharing his enthusiasm, but she looked lost again. He willed himself to slow down.

"Remember what happens after power-on reset. The program initializes the hardware and peripherals, and then reads FLASH memory to populate all the local variables in RAM before the rest of the program runs. FLASH holds the data values while the power is off. When the power comes back on, everything in RAM is garbage until it's initialized from FLASH." He stopped and crossed his arms, thinking that he'd never really considered that the problem could be due to a mismatch between the volatile and nonvolatile memory storage locations for the same variable. He'd have to remember this the next time.

He continued, "That's when the LEDs turned RED. Why?"

Josie broke eye contact. He imagined her mental wheels turning, and then prompted her again. "That's when the bad data was read into RAM for the first time."

"Are you saying that the new data isn't stored in the RAM variables like **LED_color** at the same time it's stored to FLASH?"

He shook his head. "Not at all." He drew a picture on the whiteboard behind him, starting with a square labeled NONVOLATILE FLASH and another labeled VOLATILE RAM. Inside each he printed the words "Display LED color" and drew a line to connect them, and repeated the same with Friends database variables.

Oscar pointed to the squares. "When a variable is changed, both the RAM and FLASH locations are supposed to be written at the same time, but *only for the variables that are explicitly commanded*." He tried to be as clear as possible, knowing that juggling two memory locations for the same variable could be confusing. "When the bad command to update the Friends database was executed, the 34 bytes of data were written to FLASH. But the program only commanded it to update the Friends variables, not the **LED_color** variable. So only the RAM version of the Friends database was read from the newly written FLASH and stored locally for use." He tapped on the bench for emphasis. "But there was no command to update the LED variable, so FLASH held the new corrupted value without copying it to the corresponding RAM variable. The RAM variable for **LED_color** didn't get written with the bad data until the power cycle when *everything in RAM* had to be written from scratch." He paused and waited for his explanation to sink in. "Does that make sense?"

Josie nodded. She had a serious look on her face, but he thought he could detect the light bulb warming up.

"Now, here's the plan." Oscar had some ideas about how to proceed now that he understood the problem. "While you're still thinking, I want you to fix command 0x06 in the Communicator code so that it only writes 33 bytes, as I believe it was intended to do, and then test the fix. Verify with the DEADBEEF memory-patterning function that the **LED_color** byte is no longer overwritten. I am going to find the entire set of System Test code for the Communicator and check their scripts."

Josie showed him her pad. "Wouldn't it just be this one line change to fix the last size argument and store the return status byte?"

```
status = store_data(offset, data_ptr, SIZEOF_FRIEND);
```

"Yeah. Good catch on the status byte again. Test it."

It was past midnight when they packed up to go home. Before he left, Oscar sent Randy a short email, "No software fix tonight. Already sent Mahesh home."

Oscar was already parked at his bench in the lab when Randy arrived first thing in the morning. As he made his way through the benches, Oscar could tell that he had received the late-night cryptic email.

"Mahesh told me you never called him to do a build," Randy said.

"No, we didn't."

"You've passed the deadline. I'm on my way to tell the factory to reschedule now. Is that what I should be doing?"

"I'm not changing the code. There's nothing wrong with it."

Disbelieving, Randy stared back at him.

"Really, it's under control, we're not going to pay any late fees. Josie and I figured it out." Oscar explained everything they had discovered about the System Test procedures and the mysterious command 0x06.

"This was all a strange set of circumstances, but it turns out we don't need to change the Communicator software immediately. After we discovered what caused the failure, I modified the System Test automated scripts to send the Communicator the correct commands instead of the broken 0x06 command. Eduardo is checking my changes and running the new script now. If the Communicator passes using the modified scripts, then we will have verified that the source of the problem was the bad command."

Oscar paused to gauge Randy's reaction. Randy motioned him to continue.

"What's the better news is that we don't need a new version of Communicator software. Josie fixed the broken command, but to make this deadline we are still using the one that the factory already has approved and loaded."

Randy interrupted him, "I am missing something. You're describing a change that needs to be made, so what happened to all the build and System Test time you needed before release to manufacturing?"

"Oh, that's what saves this deadline! All that time is required if we rebuild the Communicator software to fix the 0x06 command. We are not doing that. Manufacturing normally runs an abridged version of the full System Test suite that we run here. We're just going to restore a previous version of those scripts that uses the correct commands before the bad command 0x06 was introduced. We should have that completed and verified in the next few hours."

Randy dropped onto a stool and let out a long breath. A series of emotions passed over his face, from relief to pleasure to exhaustion.

Oscar quickly added, "This isn't System Test's fault, either. The bad command in our code is the source of the problem." He wanted to complain about the new custom command, but Randy was off on his own mental checklist.

Oscar waited until Randy looked up again and asked, "Are you sure? Is this all going to work?"

"Josie code-reviewed everything I changed in the System Test script. She also verified that fixing the bad 0x06 command in our software also fixes the problem, but we don't want to go that route because it would require we recompile the software and we already missed the deadline for that." Oscar finally let a smile break onto his face. "If we take the path I have proposed and update the System Test script instead, we are still on schedule. No penalties."

Randy stood and tucked his hands under his arms. "Good. Get it done."

He silently held Oscar in his gaze for several moments before he spoke again.

"I hope you realize that you can't continue to act like the lone developer any longer. You have a good reputation for firefighting the emergency problems, but you've got a team that you're not leveraging. You were promoted to technical manager and it isn't clear to me that you are making that transition effectively."

Oscar paused. He thought about Josie's involvement, and realized that her help had probably saved the deadline. Just having someone to bounce his ideas off of. He amended the thought, admitting to himself that she had been much more than that; she had understood him and her attitude inspired him to explain things. He looked up to see Randy still studying him.

Oscar shook his head, thinking that a few hours of sleep and several cups of caffeine just weren't enough to overcome the sleep deprivation he felt. Tomorrow was another day to think about what all this meant. He followed Randy out of the lab to catch another can of soda on his way back to System Test to let Eddie off the hook.

This bug was most certainly nailed.

Chapter Summary: The Case of the Irate Customer (Difficulty Level: Moderate)

In this mystery, Oscar is called in to debug an emergency problem that appeared after all testing was successfully completed and the Friend-Finder Communicator was ready to ship. The device suddenly began behaving differently, even though no software or hardware had been changed. In addition, it only failed after first power-up, right out of the box. By collaborating with Josie and with Eduardo in System Test, they find that a combination of an old software change and the payload of a new automated script used in System Test and Manufacturing causes the problem.

The Problem Symptom(s):
- Device glowed red on first power-up rather than blue as required.
- All subsequent power-ups correctly glowed blue.
- Problem only occurred after manufacturing or system test, and could not be duplicated in the lab.

Targeted Search:
- Search in the code for all places the color could be changed.
- Review of recent software changes.

The Smoking Gun:
- The color variable was the first memory location in a logical block of memory, suggesting overrun from the previous data structure.
- Changing the value of the color variable did not immediately update the LED color.

The Bug:
- An incorrect structure length was used to write data into a database. This caused a 1-byte memory overrun that corrupted the color variable stored just after the database in FLASH. This bug was masked because the corrupted FLASH value was only used to update critical variables in RAM after first power-up.

The Debugging Method Used:
- Interviewing system test engineer.
- Reviewing recent changes in the defect tracking system.
- Searching the call structure for accesses to the **LED_color** variable.
- Using watch points to allow continued code execution while waiting for a memory value to change.
- Using patterned memory (i.e., "DEADBEEF") to isolate memory writes.
- Using a team member as a sounding board to work through ideas.

The Fix:

- Changed block-memory copy command to use the correct database length (33 rather than 34 bytes).
- Used a #define rather than a hard-coded value for database length.

Verifying the Fix:

- Tested both the software fix and the system test script rollback using the patterned memory method.

Lessons Learned:

- Nonshipping software used in development and testing should be code-reviewed and placed under version control.
- Brainstorming with a team member can quickly generate a target list of ideas to explore.
- Breakpoints configured as watch points allow continued code execution.

Code Review:

The software in Figure 1-1 is pretty awful. The data are described twice: once as the **friend_data** structure and once as list of #defines, and the layouts are not identical. A coder might be tempted to simply copy into the data structure and the data would not end up in the right places. And the descriptive #defines at the top are positive, but why didn't the developer use #defines for array lengths and to clarify the confusing cases in the switch? The **process_commands()** function can be rewritten a number of ways to increase readability and maintainability. Oscar was rightfully upset at the software in Figure 1-5; the structure is accessed inappropriately and this test code does not duplicate the actual code.

What Caused the Real-World Bug? During the seven-month cruise to Mars, the Mars Expedition Rover Spirit collected science data to FLASH memory. Lots of it. Once on the planet's surface, even more data was stuffed into FLASH. But a flaw in the file system software prevented memory from being correctly deleted after the data had been sent to earth, allowing all available free memory to be quickly consumed. When more memory was requested, the system crashed. A watchdog timer faithfully rebooted the vehicle's code but on each start-up the code first tried to allocate yet more FLASH. So it rebooted again and again till the batteries nearly died, driving the system into a safe mode.

When engineers uploaded diagnostic software, they were able to free up memory, fixing this embedded application from 100 million miles away. The problem could have been caught in analysis - but wasn't. Testing, too, showed evidence that had not been investigated. Ground tests never exercised the system in "a flight-like way," certainly nowhere near the mission goal of 10 months. [1]

References

[1] Reeves, G., Neilson, T., and Litwin, T. (2004), "Mars Exploration Rover Spirit Vehicle Anomaly Report," Jet Propulsion Laboratory Document No. D-22919, May 12, 2004.

Real-World Bug [Location: Scotland] "Killer software" was once a joke term. No longer. During the 1990s certain British Chinook military helicopters had their avionics replaced with more sophisticated units running big hunks of firmware. A variety of problems appeared; sometimes the rotors started spinning unexpectedly quickly in what was soon labeled an "uncommanded run up." Pilots were instructed, naturally, to hit the reset button when such odd events occurred. In 1994, an updated Chinook went down, killing the pilots and crew.

Chapter 2

The Newest Employee: Learning the Embedded Ropes Through Code Inheritance

h, I can't BELIEVE this guy! This code is terrible!" Li Mei drew another red line through a software listing and grimaced. "I thought this part of the software already worked."

Realizing she had spoken aloud, she quickly looked around to see if anyone had been standing nearby. Talking out loud was definitely a problem with cubicles.

Her first day wasn't going as expected.

Josie had shown her around and she had met many people, including her new manager, Oscar. He had been in the middle of a big deadline and he didn't seem that excited to see her. It was nearly time to go home and her only visitor had been the human resources manager with a stack of forms. She was alone, feeling forgotten in the bustle of activity around her.

She'd thought her first day would be more exciting. It was supposed to be her special day - her very first day of work, ever, and no one cared. Did joining the Real World mean that she was just an extra person to clean up software messes?

She flipped through the stack of documentation and the Change Request form again. The title of the CR was "Fix user button function (software)" and it was a Severity 3, Priority 2 CR. The form explained that a Severity 3 bug was expected to be significant although limited to a confined section of the code, but Priority 2 meant that it needed to be fixed with some urgency. It was confusing having two numbers to describe the problem, but she reasoned that a less severe problem, like a misspelled display message, might be needed immediately, even before a terrible problem, such as out-of-memory, on a different project. So she'd better get moving on a solution.

Most of the documentation was hardware specifications for the narrow-beam ultrasonic transducer inside the Meter Magic electronic tape measure. Josie told her the hardware was working, but prototypes that included the hardware had been delayed. She was more interested in software specifications anyway, but there weren't any. She would just have to figure it out from the code.

But the code was terrible!

So far, she found this Benjamin person did not like to type very much. His variable names were too short. He didn't comment anything. The more she read, the more she thought about starting the software from scratch, but she must understand what he coded in the first place to understand what this software was supposed to do!

Oscar said something about having one week to get this working. Shouldn't she get one of these measuring things to play with? She didn't even know how to compile the code! The anxiety built as she searched her computer for any type of development tools. Would she be able to work in UNIX, or would she have to use PC-based tools? She didn't even know what this Meter Magic thing looked like!

With her throat beginning to constrict, she pushed away from the computer and willed herself to stop thinking crazy thoughts. She only knew one thing to do with this horrible software, something she had learned from a very good teaching assistant in college.

She would make a flowchart.

Placing a printout of the main routine next to her (shown in Figure 2-1), she turned to face the whiteboard that hung on her cube wall. Brand-new markers filled the tray. She picked up the red one and wrote "main" at the top of the board.

Reader Instructions: Before continuing, grab a piece of paper and sketch out a basic flowchart for the function main(). If you don't know what something does, add a box to your flowchart and assign a temporary description like "stores variable" or "performs function on variable." As you work, do you see any problems with this section of code? List anything you find.

```
void main(void)
{
    double dists[100];
    double dist;
    int i=0;
    int j;
    char c;
    char tmpstr[100];
    double ans;

    init();

    do {
        dist = measure();
        dists[i] = dist;
        if (i == 10)          i = 0;
        i++;
        sprintf(tmpstr, "%lf ft", dist);
        dmess(tmpstr);
        dmess("Meter Magic");
        c = getch();
        if (c == 'r')          i = 0;
        if (c == 'x')
        {
            ans = dists[i-2] * dists[i-3];
            sprintf(tmpstr, "%.2lf x %.2lf = %.2lf", dists[i-2], dists[i-3], ans);
            dmess(tmpstr);
        }
        if (c == 'c')
        {
            for (j=0; j<i; j++)
            {
                sprintf(tmpstr, "%d) %lf ft", i+1, dists[i]);
                dmess(tmpstr);
                getch();
            }
        }
    } while (1);
}
```

Figure 2-1 Original Software Listing.

As soon as she began, the challenge of reverse engineering the code eased her tension and the flowchart began to take shape. First, the variable declarations looked okay. Next, she trudged through the **init()** function and found that it included all the initialization software for the hardware - the ports, the LCD display, and the calibration algorithms all lumped together in one big, long routine. It looked messy, so she decided to ignore it for the moment and concentrate on the main routine.

Everything in **main()** was inside one big do-while loop that ran forever. She drew a vertical line on the software listing linking the do and while commands together so she could see where the loop started and stopped. Next, she linked each opening and closing curly brace to encapsulate the if-statement blocks of code. All were properly matched. It would be a lot easier to understand, she thought, if Benjamin had placed the opening and closing curly braces in the same column, rather than staggered like this. Then she could match them up easier.

Inside the loop, **measure()** was called before anything else. She thought that looked odd, and speculated that the distance measurement might occur in that function. A quick scan through the routine yielded no better prospects, so she added a box to her flowchart and labeled it "Take Measurement." Next, the new measurement was added to an array, `dists[]`. Distances, she thought. And it looked like only ten could be stored.

Li Mei tapped the back end of the marker against the whiteboard as she stared at the next three lines. This must print the result on a display, and she guessed that `dmess()` somehow controlled that function. She hadn't looked at the `dmess()` function yet, but the next line was a second call to `dmess()`. That was another strange thing. Wouldn't the second call just overwrite the previous measurement result on the display? Unless it somehow queued up display messages or something like that. Well, she decided, just make another entry in the flowchart and keep going.

The next line contained a `getch()`. That was logical, since a tape measure should probably receive some input from the user, and she counted three commands that the software would accept: 'r', 'x', and 'c' - whatever they meant. She drew three new decision boxes for these commands, and finished the flowchart by drawing an arrow from the bottom back up to the "Take Measurement" box.

That was it.

Li Mei sat back down, perched on the edge of her chair and stared at the flowchart (shown in Figure 2-2). She was determined to understand this code before quitting time, and quickly realized that the 'r' command was easy - the user could reset the index into the array of measurements by setting the measurement counter to zero. The 'x' command was also pretty clear; it multiplied two of the measurements together and displayed the result. Then she frowned; 'c' displayed a measurement several times in a row. She couldn't seem to crack the reason for this command and it frustrated her, but it was already after 5 o'clock and she heard people leaving for the day.

The 'c' must stand for Confusing, she decided, and solemnly noted "Print Confusing something" in her flowchart, closed down her computer and walked out of the building alone.

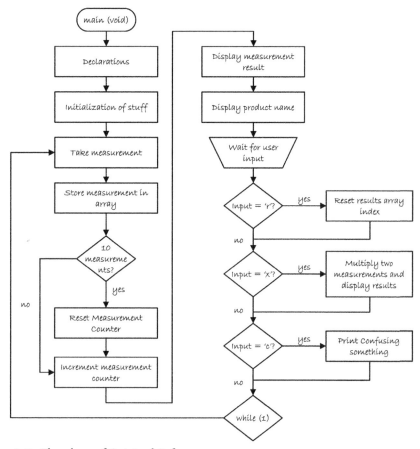

Figure 2-2 Flowchart of Original Software.

The next morning, Li Mei was well into a list of things she wanted to ask Josie. Things that she didn't understand about this software.

She'd gone home despondent and had eaten alone, while surfing the web for new-job horror stories. They left her more depressed, thinking maybe she was expecting too much. She hoped she would not be too lonely at Hudson Technologies. She would still try her best.

Around her the cubicle area was quiet. Most people were probably working in their labs. Sticking her head out into the aisle, she looked around at her new surroundings. Back at school, she had shared space with others, so she was used to concentrating amidst the comings and goings of other students and staff. This environment seemed much more formal and forbidding.

She wasn't sure what to do next with the software, so she decided to take a break and wander around. The cubicle area was large and open with street signs at major aisle intersections. Her cubicle was in the same aisle as Josie's, on Feynman Lane. She followed Feynman Lane past Hubble Avenue and Edison Avenue to the far side of the building and found herself in the marketing area. It seemed as if she had penetrated an invisible boundary; the wall color and carpeting had changed from boring beige to a deep green with maroon accents. This area had more real offices with wood doors, and a large glass-enclosed conference room with leather chairs and a large dark-wood table.

There were no more street signs.

Down the next hall she found even nicer offices. Probably the president and vice president, she supposed. She peeked into an open door and caught a glimpse of dark wood and suits as she passed, and looked backward to see a nicely dressed woman carrying folders into the conference room. She suddenly felt as if she were trespassing, and continued without slowing until she reached the end of the next aisle. Beyond a water fountain and set of restrooms, she found the back doors to the cafeteria. Breathing a small sigh of relief at the now-familiar surroundings on Hopper Avenue, she spotted Josie across the cafeteria filling her mug with coffee.

Josie waved her over. "How are you coming along?"

"Okay. I was just walking around. I like the street names." Li Mei joined her in the cashier's line.

"Isn't that cool? Navigating in the cubicle farm is a lot easier with street signs. I love that we are on Feynman Lane - Richard Feynman was an awesome guy."

"Who was he?" Li Mei considered a muffin.

"A physicist from the '50s. He worked at Los Alamos during World War II on the bomb, and then taught physics at Princeton and Cal Tech." She scooted her coffee mug up the line as she dug money from her back pocket. "There's one story where he's in the dining hall and got interested in the way that the school medallion on the edge of the plates wobbled as the plate spun. It made him think about complex motion, and he ended up with a Nobel prize. All from a dinner plate. Cool, huh?"

Josie paused to pick up a wad of singles that had dropped to the floor. "I also like that he didn't really care what other people thought of him. He ended up having great adventures based just on his curiosity about the world around him. I've got a couple books about him if you're interested."

Li Mei smiled at her as they reached the cashier, "Okay, thank you. I would like to read about this person because we are living on his street." She was interested, but she really wanted to change the subject.

She blurted, "I am having trouble with the Meter Magic. I have read everything and looked at the code. I think there might be many problems."

Josie laughed. "I don't doubt that. Benjamin wasn't a great programmer."

Li Mei looked up. "What happened to him? Did Oscar fire him?"

"No, but I think he was about to. Benjamin isolated himself and spent too much time trying to get his code to work. I don't think he really liked coding. Oscar started to lean on him for not following the coding standards and skipping code reviews, and Benjamin ended up resigning." She pocketed her change and took a drink of the coffee. "Honestly, I'm glad he's gone. If he had been a good coder, it would be different, but he didn't want to work with anyone else." She shrugged and started walking. "But now you inherit his disaster."

"I am not sure I know which parts are the disasters, so I made a flowchart to understand it. I think it's missing things."

"That is entirely possible. Did you check the documentation to see what the features are?"

"Yes, but it is not useful!" Li Mei felt her frustration start to rise again. "It's all about the hardware and the ultrasonic beam. It has no software discussion." Li Mei stuck her hands in her pockets. She knew Josie was busy with Oscar's deadline, and the tense man hanging out around the lab door was Oscar's manager, Randy.

"I was wondering if you could look at my flowchart and tell me if I understand it properly? When you are not busy?"

Josie looked at her watch, "Actually, I've got some time now." She briefly told Li Mei of the late night that she and Oscar had the night before, and the status of the Communicator project. "I am sorry that you were abandoned; let's take a look at it." Relief flooded Li Mei's face.

Before Josie had a chance to sit down completely, Li Mei was already pointing to the whiteboard flowchart and counting problems off on her fingers as she recited.

"First, I think one big problem is that the device takes a measurement immediately, before the user presses the button. And it takes measurements all the time." She handed Josie the printout of the code. "I followed this function call, **measure()**, and it turns the ultrasound transducer on, and then waits for the measurement to be completed. That function is in another file and it is written much better with comments."

Matter-of-factly she added, "Benjamin did not write that code."

Josie leaned back in Li Mei's guest chair and stared at the ceiling, amused that Li Mei was already on a first-name basis with Benjamin. With Li Mei's level of agitation, Josie wondered if she had some other names for him as well.

Josie told her, "Some of this code was written as the hardware was being developed, so that interface function should already work. Let's assume for now that it's good and just worry about the main code. Tell me why you think it takes measurements all the time. What do you mean?"

Li Mei pointed to the bottom half of the do-loop where key presses were processed. "Every time the user presses a button, the program does something and then wraps around and takes another measurement, although I guess that is mainly what the tool is supposed to do."

"No, you're right." Josie let the chair drop to all four feet and leaned forward to point at the whiteboard. "Look back at your flowchart - sometimes it's easier to see the mistakes in the software that way. If the user presses the button to reset the memory, should that cause the tool to take a measurement? That kind of defeats the purpose of clearing the memory, doesn't it? You erase memory, then BOOM! Immediately another measurement gets stored in there."

Josie thought it was a good sign that Li Mei had generated a flowchart to understand the software. But did she know how to use it effectively?

"No," Li Mei admitted, "it's not good to take measurements all the time, but there is no button to just take a measurement. So this device has to take measurements at some other times."

"No button for measurements?" Josie was taken aback, and then smacked her forehead. "It's got a button for measurements, but how the heck are you supposed to know when I didn't give you one of them to play with? Let me get one."

She left and returned a few moments later with a device about the size of a hand-held organizer and handed it to Li Mei, apologizing. "We got samples last week but no one has been available to work on the software. It's loaded with whatever software Benjamin left." Li Mei took the device and played with several of the buttons, then leaned across Josie as she attempted to measure the area of her cube. She stared at the display quizzically and then showed it to Josie, who laughed when she saw it.

"1,220 square feet! This says you've got quite a mansion here - where are you hiding the rest of it?"

Li Mei allowed a microscopic smile to escape before sliding down into her chair. "See, this software is missing a lot of lines. And it has bugs."

"It does," Josie agreed. She paused to think. "Here's what I think we should do. You made a good start creating the flowchart; now I want to see if you can use that knowledge to identify as many problems in the code as you can, before we go run the code in the lab. Then I will set you up in the lab with the debugger. Deal?"

"Okay. One thing at a time. It's a deal."

For the rest of the morning Josie challenged her to identify everything she thought was wrong with the software. As lunch time neared, Josie reviewed the list carefully printed next to Li Mei's flowchart.

- Measurement taken too often, and not on correct button press
- No way to take just a single measurement
- Display of product name at wrong times
- Command 'c' doesn't make sense
- Code for several user buttons on device not implemented in software
- Multiply command doesn't work right - indexes wrong?

Josie nodded her approval. "There are a couple of ways to approach this. On one hand, you could just fix the problems you found, and then add the missing code. On the other hand, there are some architectural issues here that you might want to correct first. Which should you do?"

Although Li Mei had passed the interview coding test, Josie wondered how extensive her experience was with different programming methods. Would she recognize that using a switch statement to process the user commands would be more readable and require less code space?

Josie hoped so. It was much more fun to work with someone who thought about better ways to solve problems, although many developers didn't think that way. They just focused on getting the software to work, regardless of what it looked like.

Li Mei had been sketching blocks on her notepad. "I think the most important thing this routine does is wait for the user to press keys, and then it should do what the command is supposed to do." She paused and brushed the hair back from her serious eyes. "I would make a new command 'm' to start a measurement. It should be a separate if-statement."

"Good start. That fixes one of the big existing problems. But what about the missing commands?"

"I would make a new if-statement for each of the missing commands."

Josie nodded. "You are on the right track. That will move all the code out of the main part of the do-while and into separate sections. However, I do see some issues with the way the if-statements are done now. Each logical is evaluated, even if the command has already been serviced by a previous if-statement."

Josie paused to see if Li Mei would suggest an alternative, but she didn't. Instead Li Mei said, "I don't understand why the if-statements are bad. If it works correctly, then why should we change it?"

"This code will work correctly, but it's not efficient and not the best solution. Think about what's being tested by the if-statements: user commands. The user can only request one command at a time. If the command is to reset the array, the command will not magically change to another command after the array is reset, so the code should not check any other options after it finds the right one."

Li Mei stared at her flowchart. "We could add else-statements. That would help."

"That's good! That would limit processing to just one of those code segments each time, but it can get messy. But I am partial to a different way of implementing user commands: using a switch statement. Do you know what that is?"

Recognition flooded Li Mei's eyes. "Oh yes, I know what that is! I can make this a switch, and then it will be easier to read, too." She started writing and quickly showed Josie her notepad.

```
switch(c) {
    case 'r':
        break;
    case 'x':
        break;
    case 'c':
        break;
    case 'm':
        break;
    default :
        break;

}
```

Josie checked her pad and smiled. "That's exactly it. How about you first rewrite this code using a switch statement and just add more empty case statements for the commands that Benjamin didn't do the first time. That'll give you a good shell to work from.

"In the meantime, I am going to update this CR to a Severity 2. The problems are much more extensive than Benjamin admitted. He didn't even finish this software and I doubt Oscar realizes this. Hiding problems just makes it harder on everyone else." Josie stood. "Give me a call when you are ready and we'll start debugging in the lab."

It was after lunch by the time Li Mei finished her first new version of the code (shown in Figure 2-3), and she was proud of the way it looked. She had replaced all of the if-statements with a switch. It was as perfect as she could make it, and she hoped it would compile the first time without any errors in front of Josie.

She crossed the fingers of one hand and shoved it under her leg as she hit the key to compile her new code with the other. A few seconds later, the screen displayed "Status: Success, Errors: 0" and she slid from her seat and turned to Josie. "See! I told you it would be perfect!"

Josie grinned. "So, make returned zero - it must work!"

"What?" Li Mei stopped short and looked doubtfully at her.

"Nothing - that's something an old friend used to say. That if the code compiled without errors, it must mean that the software would work correctly. He used to mock another guy on the team who released compiled code without testing it first."

"Oh, well, I still plan to test it."

Josie burst out laughing. "I didn't mean that you weren't! I mean, I hope you will! Come on, let's test your new code."

```
do
{
    c = getch();              /* Wait for user input   */
    switch(c)
    {
        case 'c':             /* Mystery command */
            for (j=0; j<i; j++)
            {
                sprintf(tmpstr, "%d) %lf ft", i+1, dists[i]);
                dmess(tmpstr);
                getch();
            }
            break;
        case 'm':             /* Take measurement and display results */
            dist = measure();
            dists[i] = dist;
            if (i == 10)
                i = 0;
            i++;
            sprintf(tmpstr, "%lf ft", dist);
            dmess(tmpstr);
            dmess("Meter Magic");  /* ??? */
            break;
        case 'r':             /* Reset index into memory */
            i = 0;
            break;
        case 'x':             /* Multiply last two values together */
            ans = dists[i-2] * dists[i-3];
            sprintf(tmpstr, "%.2lf x %.2lf = %.2lf", dists[i-2], dists[i-3], ans);
            dmess(tmpstr);
            break;
        case '+':             /* Add last two values together */
            break;
        case '-':             /* Subtract last two values */
            break;
        default:
            break;
    }
}
while (1);
```

Figure 2-3 New Code with Switch Program Control.

Reader Instructions: What is going to happen with the software now? "Play computer" by mentally trying different user commands and make a list of correct and incorrect behavior.

Very quickly, they ran into problems. Li Mei pressed the measurement button several times in a row and was pleased to see reasonable distance numbers being displayed, but when she tried to multiply two numbers together, the answer was wrong. She continued taking measurements and computing areas, when the device suddenly reported an area of zero square feet.

She sighed and laid the tape measure on the bench.

"You changed the behavior of the device when you created the 'm' case to take a measurement." Josie prompted her. "What did you change?"

"Nothing in the code for area calculation!"

"Yeah, but you changed something that *affects* the area calculations code. Think."

"Yes. One thing at a time." Li Mei took a deep breath and looked back at the original code printout. "If I could use a debugger, I could check the value of the two measurements, and then see if the multiply operation is correct."

"That's true, but you won't always have a debugger. Think with your brain, not with your debugger. What code no longer runs?"

And then she saw it. "I stopped the measurement code from running every time. That means the measurement counter index no longer gets incremented each time. That means it multiplies the wrong values together!"

"Great! Now tell me what two values it uses."

"Now it ignores the last measurement and multiplies together the two before that."

Josie nodded, visibly pleased. "Good detective work. I imagine Benjamin couldn't get it to work originally, and probably used the debugger to randomly change the array indices until he got the right answer. I suspected something was wrong right away because the array indices were strange. You often see paired offsets like [i-0] and [i-1], or like [i-1] and [i-2], depending on when the index counter is incremented, but indexing back to [i-2] and [i-3] is a red flag.

"Remember your palatial office space? All a result of crappy programming practices!"

Without pausing, Josie continued to grill her, "But why did the last area calculation come out as zero when you had just finished a string of valid distance measurements?"

Li Mei turned back to the code. One thing at a time. Think. They'd both taken a bunch of measurements but neither had used the reset function. Where could a zero suddenly come from? Had the measurement counter rolled over? She thought about what would happen if the counter rolled over to zero and then realized the multiply would reference **dist[-3] and dist[-2]**. She felt excitement rise in her chest; the way Josie was presenting this to her, it was just like one little puzzle after another.

She looked up to see Josie peering at her. "The index rolls over and the multiply operation uses bad values. I can fix this."

With Josie watching over her shoulder, Li Mei fixed the multiply operation, using a modulo operator, %, to properly handle the subtraction operation of the array index. Even if the distances array filled up and the index rolled over, it would still access previous measurements at the end of the array correctly. When they had finished removing the bugs in the code Benjamin had left, Josie raised one more issue.

"Before you start adding new code, I'd like you to take one more pass at this code and make the variable names more descriptive and add comments to the code." Josie stood and stretched. "Your email probably isn't working yet, but Oscar just sent around the coding standards document again. I'll send a copy to the printer in the hallway for you."

"I will make those changes and show you first thing tomorrow morning." Li Mei rose as Josie gathered her things to leave.

Today had been a much better day than yesterday, and she was happy that Josie helped her debug the software. She was going to be a team member after all.

As Josie walked out of the lab she called after her, "Thank you, Josie!"

On her third morning at Hudson Technologies, Li Mei dropped her backpack in her cube and heard a commotion coming from Josie's cube.

"The email says the manufacturing run completed without errors, and the first set of Friend-Finder Communicators powered on blue! Success!" Josie's voice floated over the cube walls, punctuated by Oscar's directive to "Rock on!"

Li Mei smiled at the excitement; they must have fixed the problem. She walked next door. "Hi. The project is successful?"

Oscar turned to face her. "Yes, Josie and I were reasonably sure everything would work out, but you never know for sure until everything plays out. Randy - that's my manager - has been pacing for days. Now, officially, that project is delivered and out of our hair."

Josie interrupted him, "Until they find bugs, and then we get to make at least one revision."

"True, but that's expected and already been negotiated. Now that ball-breaker Anders is off my freaking back. Excuse me." Oscar glanced at Li Mei but she seemed unphased. He pushed off the wall to face her.

"Li Mei, I am officially remiss in not meeting with my newest employee yet, and I would request that you join me for lunch this afternoon. Are you vegetarian?" Li Mei shook her head. "Then we shall venture to Molly's, a fine establishment for excellent burgers and dark ale among friends." Li Mei grinned at his mock-seriousness and graciously accepted his offer. He left her with instructions to meet him at quarter to noon.

After he disappeared onto Ritchie Way, Li Mei turned to Josie with amazement in her eyes.

Josie nodded. "He can be a piece of work, but he's an incredibly smart guy."

"He seemed really happy. I wasn't sure my first day."

"Well, that deadline was stressful for a lot of reasons. Benjamin quit, the project had financial penalty clauses for every day we were late, and he didn't know much about that product." Josie added, "Sometimes it gets like that, and you just have to deal with it. We go through periods of late nights and stress, but on the flip side Oscar will randomly give us time off for good behavior."

Li Mei settled into Josie's guest chair. "How long have you been here? Has Oscar always been your boss?"

"I have been here for five years and Oscar has been my manager for the last year. I worked for another manager before, but he wasn't a developer. When I came, Oscar was a senior developer and kind of a superstar debugger, so they promoted him to technical manager and I was put in his team." Josie fiddled with a pen as she spoke. "I like working for Oscar better than my old manager because he understands the technical stuff. I can ask him questions and he tries to give me real answers, not BS. But," she conceded, "he's a new manager and sometimes he gets distracted with projects and I have to bug him for information."

Li Mei sat listening with her head down. "I am nervous about having lunch with him. I hope he likes me."

"He will - just keep being honest and showing that you want to do a good job. And don't order any alcohol at lunchtime, only after work, okay?"

"Deal! Will you look at my new program right after lunch?"

"Absolutely." Josie added with a smile, "It will be perfect, right?"

Li Mei walked back to her cube after her lunch with Oscar, both energized and nervous from their conversations.

Josie had been right; Oscar was scary smart and also very demanding that she not develop any bad coding habits. After reading the original Meter Magic code, she agreed with him that documentation and communication were Good Things and she made a pledge to herself not to let Josie and Oscar down. Her first few days had been a roller coaster, but she liked Josie and hoped that she would be able to live up to Oscar's expectations.

She had successfully compiled her changes when Josie's voice startled her from deep immersion in the new code. She turned quickly to find Josie sitting comfortably with her legs crossed in Li Mei's guest chair.

"How was Molly's? Did your carnivore boss order his burger raw in the middle with a slight char on either side?"

"Josie - you scared me!" Caught by surprise, Li Mei felt her back stiffen even beyond her normally perfect posture. "You give me a heart attack!"

Josie doubled over in laughter. "I'm sorry - I didn't mean to surprise you quite that much."

Li Mei took a deep breath and gathered her thoughts. "Yes, lunch was very good. Oscar explained things to me and talked about some projects I might do. Molly's has an interesting atmosphere and was full of people who look like engineers."

She added, shyly, "Like me now."

"Yeah, it's a popular spot for the tech companies around here. Central Jersey has a ton of great restaurants of all types. We should go to Molly's after work sometime." She waved her hand. "But now, let's see the perfect code."

As Josie dipped her head to the code listing, Li Mei mentally crossed her fingers. She had changed variable names, encapsulated code, and added nicer indentation for readability.

Josie reviewed the code, stone faced, for what felt like forever before responding. "This is a vast improvement." She fell silent again, then added, "You read the coding standards, didn't you?"

Li Mei flushed and admitted, "Yes, two times. I found four violations that I fixed."

Josie snorted up her soda, trying not to laugh. "Li Mei, you crack me up. I know you are trying to make a good impression, but don't worry about solving all the problems of the world this week, okay?"

Li Mei started to protest as Josie silenced her with a wave of her hand. "Keep going. Walk me through your new code line by line and explain the changes."

When Li Mei finished, she posed a question back to Josie. "I thought about the if-else statements and the switch some more, and I really don't understand why the switch is better."

"Well, I think I mentioned that using a switch is good when the different options are discrete and have nothing to do with one another, like processing user inputs or different commands in a communication protocol." Josie paused to think. "Or to control large unrelated chunks of code in a state machine. On the other hand, if-statements are much better for continuous ranges of values."

She grabbed a pen and wrote:

```
if (x > 10 && x <= 20) {
    (stuff)
}
else if (x <= 50) {
    (other stuff)
}
else {
    (still different stuff)
}
```

"See, you can't do that easily with a switch. When you design your software, you should think about what method you want to use, switch versus if, because it will give the reader an idea of the type of processing that will occur. This is a way to make the code more self-documenting.

"In your case, I think it makes it more readable, and easier to add new case statements without introducing errors. Also, the compiler can optimize the code, which can make it smaller and faster."

Josie waited for Li Mei to nod her understanding, and then switched gears. "Do you know what we just did? That was an informal code review. And when you show me that this code compiles and runs as expected, you can go ahead and submit it."

"Okay." Li Mei stared at her, clearly not sure what was expected.

"So here's the official procedure. You made major changes to the code, so you are supposed to have a formal code review with three developers. But before that, you must have an informal review with one other person where you explain the changes you made, and you show that the code performs as expected." Josie made notes for her on a sticky pad. "We'll finish the informal code review today or tomorrow and then I'll schedule the formal review for you the first time."

"Is the code really okay?"

"If it runs correctly, but I think it's really close. You should have it finished soon. You pulled a couple of really important points from the coding standard that I should commend you on. First, by changing the variable and function names to be more descriptive, you made the software self-documenting. Very nice. You also added several #defines that make it easier to understand and maintain. The layout is also a vast improvement."

Then she leaned in conspiratorially. "I noticed you used three-space tabs. Others use four- or five-space tabs."

After a long pause, she added, "It's a war." Josie watched Li Mei's eyes go wide. "Just pick one and stick to your guns. I am a three-space coder too."

Josie brought the meeting to order. "Has everyone had a chance to review the software? Please tell me how long your preparation time was."

Li Mei twined her fingers and listened as each of the participants responded. She already knew Ravi, but she had never met Bob or Sundara, and she gave a small smile to each of them. Too soon, Josie called on her to begin explaining the code changes, so she took a deep breath and began.

> **Reader Instructions:** Review Li Mei's final code (shown in Figure 2-4) and make a list of each improvement that she made. While together her changes make the Meter Magic a properly functioning device, identify the secondary benefits for each of her improvements (e.g., improved readability, maintainability).

"This Meter Magic didn't have any software requirements document, so Josie and Mike from marketing helped me make one from other documents and Oscar approved it. We will use that for this code review. I started with the first code listing, which ran but did not work. The final code listing has several changes to fix problems and satisfy requirements. Before I review the code line for line, I would like to give a list of overall changes I made."

Li Mei switched papers and read, "I changed all the variables to more descriptive names and added comments to explain the code. The main loop was originally controlled with many individual if-statements, which I replaced with a switch statement to control user inputs. I made some new functions to encapsulate some code, and also to add error checking. Some arrays could be accessed out of bounds, like the measurement array and also the display string array. I also added white space and formatting for easy reading." She looked around the table as she added, "I hope it was easy to read." To her relief, she saw heads nod, and she continued to the final code listing (shown in Figure 2-4).

```
#define MAX_NUM_MEASUREMENTS    100
#define MAX_LEN                 100
#define MEASUREMENT_TYPE        1
#define AREA_TYPE               2
#define ERROR_TYPE              3

void main(void)
{
    /* Variables for distance and area measurements      */
    double  distances[MAX_NUM_MEASUREMENTS];  /* Storage array          */
    double  area;                   /* Area calculation result          */
    double  meas1, meas2;           /* Temporary variables for calculations */
    int     meas_ctr = 0;           /* Counter for distance measurements   */
    char    user_cmmd;              /* Input from user from buttons        */
    char    tmpstr[MAX_LEN];        /* Temporary string to display results */
    int     multiply_enable_f = NO; /* Do we have 2 values to multipy?  */

    initialization_hardware();  /* Initialization of ports, beam, etc.  */
    clear_measurement_memory(distances);
    display_message(TEXT_TYPE, "Meter Magic");

    /* While the device is on, process user commands          */

    do
    {
        user_cmmd = getch();     /* Wait for input from user via buttons */
        switch(user_cmmd)
        {
            case 'c':  /* Display previously stored values one at a time */
                cycle_stored_values(meas_ctr, distances);
                break;

            case 'm':  /* Perform a measurement, store the result and display it */
                distances[meas_ctr] = measure();
                sprintf(tmpstr, "%.2lf", distances[meas_ctr]);
                display_message(MEASUREMENT_TYPE, tmpstr);
                meas_ctr = (meas_ctr+1)%MAX_NUM_MEASUREMENTS;
                if (meas_ctr >= 2)
                    multiply_enable_f = YES;
                break;

            case 'r':  /* Reset memory by clearing out values and counter */
                meas_ctr = 0;
                multiply_enable_f = NO;
                clear_measurement_memory(distances);
                break;

            case 'x':  /* Calculate area using last 2 measurement values */
                if (multiply_enable_f == YES)
                {
                    meas1 = distances[(meas_ctr-1)%MAX_NUM_MEASUREMENTS];
                    meas2 = distances[(meas_ctr-2)%MAX_NUM_MEASUREMENTS];
                    area =  meas1 * meas2;
                    sprintf(tmpstr, "%.2lf x %.2lf = %.2lf", meas1, meas2, area);
                    display_message(AREA_TYPE, tmpstr);
                }
                else
                {
                    display_message(ERROR_TYPE, "Error: Area needs 2 values");
                }
                break;
            case '+':    [READER: code for remaining functions removed for space]
                break;
            case '-':
                break;
            default:
                break;
        }
    }
    while (1);
}
```

Figure 2-4 Final Code Listing.

Nearly an hour later, the code review was complete and developers filtered out of the room. Li Mei gathered software listings as Josie finished jotting notes and rose to face her.

"Congratulations! You have officially survived your first code review in the Real World. You did good, Li Mei. How do you feel?"

Li Mei's face broke into a smile, tinged with relief. "It was not as bad as I expected. I thought everyone would not like my variable names or they would tell me to change everything because I am not experienced yet."

"Nah, the focus of a code review is to verify that the code satisfies the requirements and that your code is clean, logical, and follows coding standards. If someone sees a better way to fix a bug or if they find an error, they can make suggestions but we all decide together whether to ask you to change your implementation. Other than that, you can develop your own coding style."

Josie gathered up discarded notes and listings as they talked. "Now, to finish things off, you need to make the small changes to the software that the group suggested. I will enter the results of our review into the defect-tracking system and call the review a success pending your updates."

Li Mei nodded.

"Then, you check your new changes into the source code control system like you did before the review, and I will review your changes. Then we can start an official build of the final software and System Test will begin testing it. For now, you are done."

"Good. I am happy to finish my first software project." Li Mei stopped suddenly to face Josie. "Josie, I was so worried my first couple of days here. I didn't think anyone liked me and that I had to work quietly all by myself." She shifted the papers in her arms. "But since you started helping me, it has been so much better. Thank you so much for helping me. But will we keep working together? Or do I have to work alone on the next project?"

"Well, each project is different and on some you will be the only developer. But most projects have more than one embedded developer, and you'll also meet folks from hardware and System Test and they will become your allies, too."

The tension Li Mei was holding in her shoulders eased slightly.

"It's like this," Josie explained. "We all try to help one another, and if you have a problem you come to me or Ravi or Oscar. And don't feel nervous about it, because we are more successful as a team when we communicate. We don't want another Benjamin. But I don't think you've got it in you to be a Benjamin."

With a smile on her face, Josie stuck out her hand to Li Mei. "Welcome to the team. You're gonna do great things here."

After Li Mei left the code review room, Josie made a final pass and found a note-book sitting on one of the chairs. She quickly located her own amid her stack of listings and comments, and realized it probably belonged to one of the engineers at the review.

Looking for a name, she flipped the front cover and the notebook opened to a page in Li Mei's clean and precise handwriting.

Specific Symptoms and Bugs

- *Suspicious indexes into arrays often cause off-by-one errors.*
- *When something works for a while before breaking, suspect memory problems like boundary condition violations (accessing outside array limits).*

General Guidelines

- *Use programming elements that are appropriate for the function (e.g., switch for unrelated discrete items and if-statements for continuous variables).*
- *Make code self-documenting with descriptive names, #defines, and comments.*
- *Sometimes the comments are wrong.*
- *Consistent tab spacing and white space make code more readable.*
- *Focus on understanding main() first - don't get lost in the details.*
- *Use visual aids like flowcharts and graphs to show functional elements (blocks) and program control logic (connectors). This reveals what the program does, and identifies missing logical and functional elements.*
- *Coding standards are a good source of bug types and causes, and they also provide fixes!*
- *Think with your brain, not your debugger.*
- *Just because it compiles, doesn't mean it works.*

Chapter Summary: The Case of Inherited Code (Difficulty Level: Easier)

Li Mei inherits incomplete and undocumented software for the Meter Magic electronic tape measure; she must reverse-engineer it to understand what the product is supposed to do, and figure out what is wrong with the current implementation. Josie helps her methodically understand, re-architect, and fix the software. She also learns about the code-review process, coding standards, and the importance of documentation.

The Problem Symptom(s):
- Code was not complete; code for some buttons was missing.
- Some code did not work correctly (e.g., multiply was broken).
- Measurements inappropriately occurred for all button presses.
- The product name was displayed at the wrong time.

Targeted Search:
- Divided the software into logical chunks using curly braces to encapsulate functional blocks. Matching braces revealed program flaws.
- Created a flowchart to understand when each function processed.
- Listed many problems from the flowchart and the software listing.

The Smoking Gun:
When the initial software changes caused a square-footage calculation to return zero, she realized that the entire structure of the main routine must be addressed.

The Bugs:
Several software bugs were found.
- Array index off-by-one caused incorrect area calculations.
- Bad program control caused a new measurement on the display to be immediately overwritten with the "Meter Magic" product name, and caused processing to be performed on previously stored measurements.
- Several functions not coded at all.

The Debugging Method Used:
- Using teammates as sounding boards for ideas.
- Reformatting the code and matching curly braces.
- Clarifying functionality by using flowcharts.
- Reproducing the errors with incremental code changes.

The Fix:
The software was fixed and quality improved with these changes.
- Used a switch statement rather than a series of if-statements.
- Added comments, descriptive names, and #defines for improved readability and maintainability.
- Added error checking to ensure arrays were accessed correctly.
- Updated the display function to accept different types of display messages.

Verifying the Fix:

- Developed requirements specification (because it didn't exist) and used it to understand and then verify the needed functionality.
- Verified array boundary conditions even when memory was full.
- Created Test Plan (in consultation with System Test) for new official load of code.

Lessons Learned:

- Don't assume software that compiles works correctly.
- Even if the final listing is longer, its consistent format and structure make it easier for anyone to quickly understand and make changes.

Code Review:

The software in Figure 2-1 is a good example of why opening and closing curly braces in the same column can increase readability of code, independent of other changes. In properly formatted code, Li Mei's handwritten lines would simply be vertical lines connecting the braces. In addition, it is unusual to have the input keys coded as ASCII characters, although this can make testing easier because the code can be tested on a PC using a getc()-type interface before hardware is available.

What Caused the Real-World Bug? After the Chinook helicopter with the upgraded avionics went down, an inquiry was launched. The review board examined the software more closely, running a formal code inspection. The code was so awful they gave up after inspecting a mere 18%, which identified 486 errors, some 50 of which could lead to the loss of the machine.

The lesson: inspections find errors efficiently. The tragedy: the software community has known this since 1976, yet Chinook avionics contractors chose to ignore those lessons. [1]

References

[1] Chinook ZD 576-Report, House of Lords Reports, The United Kingdom Parliament, Session 2001-02, 31 January 2002. Accessed from http://www.publications.parliament.uk/pa/ld200102/ldselect/ldchin/25/2501.htm.

Additional Reading

Ganssle, J. (2004), *The Art of Designing Embedded Systems*, Newnes.

Chapter **3**

It Compiles with No Errors: It Must Work! Integrating Changes in a Larger System

"**H**eads up, Ravi, I broke your code." Eduardo caught up with Ravi just as he turned into the lab. "Wasn't hard, either. All I had to do was run it!" He followed Ravi in, needling him about how much he enjoyed the endless challenge of breaking developers' software.

"Gonna write a Change Request against it and the subject line will be, 'The amazing two-second-long program - hardware doesn't even get a chance to power up before it's all over!'"

Ravi turned back to glare at him and Eddie skidded to a halt, the smile melting from his face. Through gritted teeth Ravi informed him, "Defect tracking system won't accept a subject line that long. Nice try."

Eduardo tossed his hands in the air. "Yo, it's okay, man, just kidding. I wouldn't really write that in a CR."

"Eddie, I fixed that software just like they asked me to. All they wanted was someone to port the software from some ancient hardware platform to something newer, and they wanted it done, like, instantly. I ported it; the job is done, okay? If something else is broken, it's not my fault."

"Sure, no problem." Eduardo fiddled with the antenna on his cell phone as he watched Ravi stalk away. It wasn't like Ravi to be such a jerk, but maybe it was just more girlfriend problems. He shrugged and walked back to the lab to write up the test results for Ravi's code changes.

Eduardo had been a developer for almost a year before he did a stint in the System Test group, only to find that he enjoyed testing software and trying to break it. It was much more fun than writing software from scratch or fixing other developers' mistakes. Some developers seemed to look down on the people in System Test, as if they didn't have enough skills to be good coders and ended up in ST by default, but he found the experience altogether different.

You had to be clever to be an ST God, as he was.

Eduardo smiled to himself at some of the bizarre bugs he had identified. Any slacker could run a test script and print out automated results, but it took talent to decipher the results and then run additional manual tests to figure out why the test failed. Eduardo prided himself in taking the extra step. Rather than reporting the symptom of the failure, he tried to identify why the failure occurred to help the software folks quickly find the error. Ravi was usually a good guy, and Eduardo decided to spend a little more time understanding the bug for him.

Eduardo sauntered back to his bench in the lab and prepared to run more tests on Ravi's code. For this change, he had no automated test scripts - only the CR that indicated that the software program must still run correctly on newer hardware.

When he was ready, he powered on the laptop-sized device and, after a short delay, names and faces of several wild animals appeared around the perimeter of the display. A strange howling noise started, but almost immediately the display blanked and the device powered off.

Strange. He flipped the device over and looked on the back and sides. Wasn't very clear what this thing was supposed to do.

The original CR was vague, but he gathered it was an informational interface for nature preserves or wild animal parks. Apparently the old devices weren't very rugged and a series of severe weather events had cracked the cases. He thought visitors were supposed to touch a picture of an animal to learn more about that animal but, unfortunately, they didn't get much time to do it before the device shut down.

Eduardo power-cycled a unit and leaned over the table to get ready. When the program started, he immediately jabbed a spot on the screen where he had seen an animal's face appear the last time. Almost immediately, a lion's face was displayed followed by 11 other animals' faces in a ring, and this time they didn't go away. Perched, unmoving, he waited for something to happen, and was quickly rewarded by the roar of a lion, followed by a monkey's howl, the trumpeting of an elephant, some bizarre rumbling noise he'd never heard before, and then a rush of other animal calls until all twelve animals had introduced themselves.

He leaned back and smiled. Cool.

As he watched, the screen cleared and was replaced by a wide-angle shot of an open savanna with a description of the lion's habitat and eating habits, reproduction cycle, and lifespan. Selecting each of these buttons led him to other screens that seemed to work correctly.

By now, Eduardo was lost in the challenge. This was the good stuff, he thought, trying to crack the code. Sometimes he could even give hints to the developer about what might be wrong with the software, like maybe a loop counter was off by one, or that a variable was uninitialized because some repeating behavior was different the first time only. All without actually looking at the code!

He thought the initial ring of animals was an opening screen that was probably supposed to stay there so the visitor could look at each animal and choose one to learn about. He pondered, while mindlessly popping his cell-phone antenna in and out of its holder. Could he select each animal individually, even though that first screen seemed messed up? He restarted the program and jabbed along a different side of the screen, but was still presented with all 12 animals and all 12 voices, starting with the lion. But this time he was rewarded with the habitat and information for the lemur, a funky little creature with a long black-and-white striped tail. Soon, with patience, he was able to access the habitat of all 12 animals, from the mighty lion through the elusive grey wolf.

He pushed back, ready to add more information to the CR, starting with a descriptive title: "Opening screen doesn't wait long enough for user input before ending program." In the description section, he outlined the tests he had performed. He added his idea that the program also had a timing problem because all the animal's faces were displayed within a second or two, but it took a lot longer for the voices to play. Was it possible the software wasn't playing the animal sounds soon enough? Well, that was all he could do to characterize the problem without looking inside the box. Basic black-box testing.

He finished up the CR by filling in the preset fields. The problem was a Severity 1 problem, the most critical because the software was unusable. It was up to management to assign the priority level, so he left the default value of 3. If Eddie was right and they needed this change quickly, they would bump up the priority level to a 2 or a 1. The CR committee would review the new CR within the next 24 hours; most likely, it would be assigned back to Ravi to fix. He hoped his extra work would help Ravi out; he liked him and didn't want to see him get into trouble for being sloppy on this one.

Ravi wondered how long before the CR was assigned back to him.

Eduardo was right. He hadn't checked the software fully. Actually, he hadn't checked it at all. He was trying to make a deadline for Oscar, and just didn't have enough time to finish it right.

But he was sure he ported the hardware code correctly!

He logged into the defect-tracking system and called up a list of open CRs assigned to him. The new CR wasn't there yet, so he returned to his email to find another message from home about his marriage prospects. His stomach clenched into a knot as he closed the email and pushed back from his desk to clear his head.

He was about to get up when he heard a new voice behind him.

It was Li Mei. He smiled, thinking about her prior threat to come bother him for absolutely no reason at all, and he turned to greet her.

"Hi, Li Mei! Are you here to make trouble as you promised me?"

Li Mei rushed into his cube and plopped into his guest chair, clearly excited about something. "Hi, Ravi, how are you? Do you have a minute? Am I bothering you?" She looked at him expectantly, guilelessly. He found himself warming to the idea of a distracting conversation with someone new.

"Okay, Li Mei. You are very happy today; what's going on?"

She fairly bubbled over with enthusiasm. "I finished my first code review! They said my code was okay and Josie is going to approve it. Well," she amended, "I have to make two small changes, but it will be no problem." She leaned in to whisper. "I was nervous because Oscar said I had only one week to finish this, and it was not very good code." Pushing back, she added, "I'm so happy to be on this team. Josie helped me figure everything out. She is very smart, don't you think so?"

Ravi stared at her, nodding for her to continue as he felt his throat constrict over his own experience so far. Oscar was a terse manager. People make mistakes, he thought, and Oscar could cut some slack.

"Sure, yeah. Josie's pretty smart. It's good you get along with her. Oscar likes her."

"I wanted to thank you for helping me my first couple of days; you know, taking me around to get my badge programmed." She made a face, "But my picture is pretty terrible."

They shared a laugh, and Ravi wondered what to make of her.

"I am glad to help. A first day is hard, and this place can be harder."

"Why?"

"Well, Hudson Technologies does a lot of different types of projects, and sometimes we're supposed to act like consultants and go on the road. There are not many really tight teams that work together on the same project for a long time. So, it can be harder to trust people." He watched the look on her face change to concern. "But if you get on a good team and get good projects, then . . ."

"Do you like working here? On this team?"

"Well, it's okay. I mean, I've only been here six months." Then he stopped and mentally reset himself before continuing, "Listen, this place is great to learn new things, and you should get as much from it as you can. That's what I am trying to do."

Li Mei sat silently; she was watching the emotions play over his face. She had come here happy with her code review and he had deflated her happiness. Li Mei stood up to go and he jumped up to stop her from leaving.

"Hey, I didn't mean to scare you off. I am really happy that your code review went well." He put a solid convincing smile on his face and continued, "You can come by any time to share your good news, okay?"

"Yes, Ravi, I would be happy to talk to you about my next project. Josie has also told me that sometimes the developers go to . . ." she paused, and looked at the ceiling as she intoned from memory, ". . . Molly's Irish Pub and Grill, the home of excellent burgers and dark ale among friends."

Satisfied with her recitation, she looked back at him and added, "Will our team be going there soon?" She burst out laughing before he had a chance to respond. "Oscar took me for a burger, and I would like to go back and try 'a dark ale among friends.'"

"Yes," he responded with some hesitation, "Maybe we should go out sometime."

"What were you thinking?! This was a high-priority change, and you didn't even bother to run the software on the new hardware?" Oscar paused, shaking his head, and then continued somewhat less vociferously.

"Again, I ask, what were you thinking?"

Ravi balled his fists and waited. Oscar's rare outburst, complete with hand waving, had quickly come to a silent, pregnant pause.

"Well, I knew it was a high priority . . ."

"Which is an excuse to submit crap, only faster?"

"Hey," Ravi shouted back, "I said I screwed up and that I am on my way to the lab right now to fix it." Grabbing his notebook and pen, he angled out of his cube and headed to the lab, only to realize that Oscar was following him. To his mortification, Oscar trailed him to his lab bench and actually pulled up a stool beside him. Was Oscar actually going to look over his shoulder while he fixed the software?

"Okay, this needs to be fixed immediately. Where do you start?"

Clearly, Oscar was staying. Ravi pulled up the software on his computer screen and stared at it. Eduardo had said something about it running and then halting prematurely. Where *should* he start?

"Well, I will check the changes I made first."

"Tell me what you changed."

"This thing is running on a new hardware platform with a new processor and new LCD display. Mike from marketing told me that it's already a working product; it just needed the new hardware." Ravi dug through the drawer, pulled out a handful of spec sheets and dropped them on the table. "I read the hardware specifications for the new microprocessor and display, and changed the lowest level of software, the hardware device-driver interface."

Having Oscar stare at him made him nervous, so he continued talking out loud.

"I started with the digital I/O ports." As he talked, he walked a finger down each line of code. "There are five buttons on the case that are digital inputs. Things like up and down buttons to make the display brighter and darker. In the old software, they came into the processor on Port D, but that is reserved for the analog-to-digital converter subsystem, so I moved those five inputs to Port A." Oscar nodded for him to continue, and he relaxed microscopically. "So in the interface header file, I changed the #defines for the five buttons from pins on Port D to the right pins on Port A."

"How did you know which were the right pins and ports to use?"

"I got the design documents and new hardware schematics from Tom. His hardware group designed the new interface board. All the hardware input and output signals were labeled, so I knew which port lines they were coming in on."

"Good," Oscar said. "And you updated the entire interface?"

"Yes, I think so." He handed the hardware schematic to Oscar, who noted that it was fairly well marked up with check marks and Xs. "I updated the interface software for each hardware line coming to the processor and display, and then I crossed it off the schematic." He felt his confidence returning; he had gone over the changes several times. But something was missing because the device didn't work.

Oscar finished looking at the schematic and placed it carefully back on the bench.

"I am relatively confident that you ported the hardware correctly. You are generally a strong coder. But my concern is your tunnel vision. Your software and changes are part of a larger system."

Ravi interrupted him, "But I'm fixing that now!"

"I just had the CR assigned back to you. Pull it up."

Ravi logged into the system. Sure enough, a new Severity 1 Priority 2 CR blinked at the top of his list. Feeling self-conscious, he started to read Eduardo's analysis and was surprised to learn what this product did. He'd only known it was something for a wildlife preserve. Eddie's descriptions were pretty thorough.

"It looks like Eduardo did a bang-up job trying to characterize this problem for you. Don't blow off his work or the debugging information he created for you."

Ravi scrolled through the test results. From the corner of his eye, he saw Oscar steeple his fingers.

Shortly, Oscar stood up. "Keep reading and figure out where to start. I need a bio break. I'll be back."

Oscar made his way through the cubical farm to the men's room, relieved himself and stood staring into the mirror in a distracted fog. He'd avoided confronting Ravi most of the morning, debating with himself about leaving it until tomorrow. All morning, niggling resentment at his new managerial position had finally driven him into the lab to get the confrontation with Ravi over with.

Was it always this hard? Becoming a technical manager brought responsibilities that he hadn't expected. New annoyances. He knew he'd waste more time in meetings, but he hadn't expected the people issues. And he still had technical assignments, but apparently the promotion hadn't warranted an office door. Pulling himself away from the tired face looking back at him, he dried his hands and tossed the crumpled paper into the trash bin.

Walking back to the lab, he suddenly got an idea and detoured to his cubicle. He slid into his chair and pulled a palm-sized external data-storage device from his briefcase, connected it to his computer and launched a search screen. Quickly finding what he wanted, he copied a folder to a memory stick dangling from a spare port and then removed it.

Back in the hallway, he thought again about how Ravi had approached his assignment, only this time he wondered how he could entice Ravi to recognize the error Oscar thought he was making. Without his being told. He was slow to recognize his name being repeated.

"Earth to Oscar, come in!"

He turned to see Josie already walking beside him, and marveled at the large bagel she waved at him.

"Whoa, nice bagel." He tilted his head in apology. "Sorry, lost in thoughts."

"No problem." Josie took a long pull from an oversized coffee cup and added, "Hardware brought in bagels this morning, if you want one. Look at the size of this monster. I won't be needing lunch today!"

He nodded in agreement. "Hey, thanks for taking Li Mei under your wing the last week or so. Is she settling in okay?"

Josie nodded. "I think so. She just jumps right into anything, even though she doesn't have much experience yet. She'll be done with the Meter Magic project in a day or so. What are you going to have her do next?"

"I'm not sure yet. I am in the middle of getting Ravi's project back on track."

Josie was silent, and then offered, "Yeah, I heard there was a commotion in the cubes. Is everything okay?"

He hesitated and then the words tumbled out before he realized he was committing himself. "This technical manager thing is not exactly what I expected. I gave Ravi an assignment with a deadline, and he submitted crap and he knew it. I want to trust that people will do quality work - I mean, that's why they're getting paid. I shouldn't be getting blindsided by my team, right?" Clenching the memory stick, he punctuated his words with chopped motions as his worries poured out.

"Take you, for example; I give you an assignment and you get it done on time. Your work is good, thorough, and well documented. I trust you." Oscar relayed the morning's events to her, oblivious to the pleased look that appeared on her face at his spontaneous praise.

As his rant wound down he finally admitted, "Yes, I did lay into him. I should have been more private about it, but he still deserved to get called on the carpet." He finally saw her expression and it stopped him short.

"Why are you smiling?"

"You said you liked my work."

"Yes, I do. I don't have to worry about you - you're one of the good ones." Oscar was flummoxed. "You didn't know that?"

"Well, I try my best and I really like doing this kind of work." She shrugged. "It's just really good to hear it from you. That means something to me. Thanks for telling me."

Oscar looked down at the memory stick in his hand, and then slipped it into his pocket as he thought about what Josie had said. Did she really not know how he felt about her work? He tried to think back to the last time he might have told her and realized that he couldn't recall when that might have been. Here was another management responsibility that he had completely missed, assuming that everything was rolling along okay with her.

"Listen, I'm sorry, Josie. I should have given you positive feedback before now. Let's sit down soon and talk about your work and maybe what you want to work on next, okay?" He saw her nod, smile still firmly in place, and he continued, "Listen, let me get back to Ravi. I have to teach him some debugging skills. As Randy is growing fond of grilling into me lately, I don't have time to do everything myself."

Oscar made his way back across the crowded lab and slid back onto the bench stool. "Okay, here's what we're going to do. We'll solve this together so I can get an idea how you approach debugging a problem."

Ravi turned to stare at him, surprised that Oscar's voice no longer carried any anger.

"Like I said, the way you handled this assignment was not appropriate and I don't want it to happen again. But I should not have yelled at you in public. I apologize."

Oscar turned to face Ravi and got comfortable on the stool. "So, as engineers, we tend to want to work in a vacuum, but we miss vital information that can help us do our jobs better. If you talk to people on your team about what's going on in a project, sometimes they have seen the problem before or have ideas how to approach a solution."

Ravi admitted, "I'm not sure what you mean." He tapped his fingers nervously under the front edge of the stool.

"I am drawing some conclusions from Eduardo's report. Conclusions that haven't dawned on you yet, but should." Oscar paused and looked at him, exasperated. "Think about it, Ravi - Eddie could not have gotten as far in the testing as he did if your porting job was bad. In fact, I think the bug has nothing to do with mapping the hardware correctly. It is highly likely that the work you completed was done perfectly."

Had he heard Oscar correctly? Was he really praising his work? Ravi searched Oscar's face for some sign of trust.

"Okay." He gave a small nod.

Events were taking a surprising turn.

"Let's start at the beginning." Oscar pulled up the CR on the screen. "The biggest thing holding you back is your narrow focus on this assignment. The CR requests a port to the new hardware and then states: 'verify that previous operation is maintained.' That's a big red flag - make sure you validate the ENTIRE system, not just the small part you think you changed.

"Now Eduardo has already done some of this testing for you. With that in mind, let's start over. What do you do first?"

Ravi thought about Eddie's test results. "I think the opening screen has a timing problem. So we look at that code first." Ravi pulled up the code on his monitor and looked at the main loop (shown in Figure 3-1).

> **Reader Instructions:** Based on Eduardo's analysis report and Oscar's feedback, where is the problem? Only main() is shown. Focus on the big picture first, and then list any potential problems you see.

Ravi took a deep breath and tried to concentrate. He quickly scanned the initialization functions that he'd worked on before and returned to the main loop. After a function clears the display, the lion is displayed, and then all the animal sounds are queued up to be played. Next, the other animals are displayed one at a time.

Eduardo had joked with him that the program finished very quickly, but his later tests confirmed that the pictures were actually displayed. From the listing, though, he thought animal sounds should also have been playing.

Oscar's voice brought him back. "Tell me what you are thinking."

"Well, the pictures are displayed, which is good, but the sounds didn't play correctly. So, one thing I would consider is looking at the sound function and how I ported the sound hardware."

Before he could continue, Oscar cautioned him. "Solve the easy problems first, because sometimes those bugs cause side effects."

"Okay. Eddie said the program ended very quickly, so I would look for normal and abnormal ways to exit the program. Just from the main routine, I could guess that the variable `selected_animal` wasn't initialized correctly. There must be an interrupt or something that processes incoming touch screen presses - that's how the variable gets a valid value the first time. If the user doesn't touch the screen by the time all the animal faces are displayed, that variable could be garbage. If it were already set to `QUIT`, the program would end immediately."

"Okay, but how did it work correctly in the first place?"

Ravi considered a moment. "This toolset automatically sets all uninitialized variables to zero, which happens to be the `enum` value for `QUIT`. The tools I used in

```
enum { QUIT, LION, HOWLER_MONKEY, AFRICAN_ELEPHANT, EGYPTIAN_COBRA,
       POLAR_BEAR, ZEBRA, OCELOT, HUMPBACK_WHALE, RING_TAILED_LEMUR,
       GALAPAGOS_TORTOISE, ORANGUTAN, GREY_WOLF };
enum { NO, YES };
int touch_screen_input_f = NO;

void main(void)
{
    int selected_animal;

    initialize_ports_and_interrupts();
    initialize_LCD_display();
    initialize_sound_system();

    do
    {
        /* Present all the animals in ring around display edges */
        clear_display();
        display_picture(LION);
        queue_all_animal_sounds();

        display_delay(198);
        display_picture(HOWLER_MONKEY);
        display_delay(198);
        display_picture(AFRICAN_ELEPHANT);
        display_delay(198);
        display_picture(EGYPTIAN_COBRA);
        display_delay(198);
        display_picture(POLAR_BEAR);
        display_delay(198);
        display_picture(ZEBRA);
        display_delay(198);
        display_picture(OCELOT);
        display_delay(198);
        display_picture(HUMPBACK_WHALE);
        display_delay(198);
        display_picture(RING_TAILED_LEMUR);
        display_delay(198);
        display_picture(GALAPAGOS_TORTOISE);
        display_delay(198);
        display_picture(ORANGUTAN);
        display_delay(198);
        display_picture(GREY_WOLF);
        display_instruction("Select Animal to Learn About");

        if (touch_screen_input_f == YES)
            selected_animal = decode_keypress();

        /* Skip if user selects quit */
        if (selected_animal != QUIT)    {
            if (touch_screen_input_f == NO)    {
                wait_user_keypress();  /* wait 'til they select something */
                selected_animal = decode_keypress();
            }
            /* Display the animal-specific information and wait for user to return */
            display_animal_information_screens(selected_animal);
            touch_screen_input_f = NO;
        }
    } while (selected_animal != QUIT);
}
```

Figure 3-1 Original Software Listing for main().

school only set uninitialized global variables to zero, so it's possible the old software worked by accident."

"Not bad. Actually, a pretty good analysis with your manager breathing down your neck." Ravi turned to see a small grin on Oscar's face. He wondered how much of this was predetermined, and blurted out his suspicion.

"Do you already know what the bug is?"

The question caught Oscar off guard, but he admitted that he might.

"How do you know?"

Oscar looked down at the bench before he answered. "Well, that's a legitimate question. I could prompt you so you could figure it out yourself, or I could tell you. But if I just told you, what would you have learned?"

Raising his gaze to meet Ravi's eyes, he continued, "And what good would you be on the team if I had to debug all your assignments for you?" Oscar appeared to be uncomfortable with the conversation, and turned his head away to scratch at his neck.

"Well, tell me what to look for next so I can figure it out. Do I look at the sound function?"

Oscar's face softened almost mischievously, as he stood to extract the memory stick from his pocket. "Hold off on that for now and deal with the main loop. I want to load a program on a computer to show you."

By the time Oscar returned, Ravi had already fixed the initialization problem by adding a new **NO_ANIMAL_SELECTED_YET** enum, and he verified that the program no longer ended instantaneously. All the animal faces quickly ringed the display and the animal sounds started to play one after the other, but he'd identified the next problem. The animal sounds didn't stop when he selected one animal from the menu. Other animal sounds continued to play until all 12 were finished.

Oscar called out, waving Ravi over. Then he launched a simple DOS window as he talked.

"I want to show you a program that has the same bug as your program. Well," he amended, "'bug' is not exactly the right description. I should say that it suffers from the same *problem* that yours does."

Oscar turned back to the screen and hit return. The DOS window turned black and two vertical white bars, perhaps an inch tall, appeared near the left and right sides of the screen. A white ball was positioned in the middle. Oscar used the cursor keys and the right paddle moved up and down, zipping from the top to the bottom of the screen almost instantly. He readied himself and then hit the space bar. The ball shot like a bullet to the left, bounced off the left paddle and sped off at an angle to the right. Oscar's attempt to move the right paddle to hit the ball proved humorously futile. After two more near instantaneous failures, the window displayed GAME OVER.

"That," Oscar announced, "was the world's shortest game of Pong. Ever heard of Pong?"

"Yes, that's a very old computer game. I've not played it."

"It's the oldest. Want to play it now?" Oscar motioned to the keyboard, not surprised to see Ravi shake his head.

"It's too fast! There is no way to hit the ball. Can you get a better version that works right? Where did you get it?"

"I wrote it."

Ravi tried to hide his grin at the thought that Oscar had made a coding mistake.

Oscar continued, "I wrote this program in 1992. And it ran perfectly. My high score is over 10,000."

Oscar caught Ravi's eyes and held them. "So why doesn't it work now?"

"Think about the similarities between your project and my Pong game," Oscar instructed. "Yours is working software that got a hardware upgrade. The upgrade essentially *broke* the system because it wasn't compatible, but that's typical for embedded systems because they are so customized to the application. My Pong game is software that also worked correctly on my top-of-the-line 33-MHz 486DX desktop computer, but it doesn't work right now. To run it on this computer, I would have to make software changes too."

Oscar found himself warming to the discussion, reminiscing about college and the radical changes in computing power since then. Was he practically handing Ravi the answer? They were back at Ravi's bench, having reviewed the limited software listings a second time.

To Ravi's frustration, Oscar wouldn't let him dig any further into the code.

Ravi sat, unmoving. "Both programs play too fast. The Pong ball moves too fast, and the animal faces are displayed too fast. Each animal's face is displayed faster than the animal sound."

"Excellent. Root cause?"

"There's a software delay between displaying each animal picture. Those delays aren't long enough anymore. We could make that 198 number bigger and see what happens."

Oscar nodded. "Yes, we can change that number, but first tell me why you think that will work. Randomly making changes is not the sign of a good debugger."

"Well, that's the only thing between each line to display an animal's face."

"Not good enough. Let's first check what that function does. If that 198 value is in units of absolute time, like milliseconds, the function should run the same on old or new hardware." Oscar paused as Ravi searched for the function (shown in Figure 3-2).

```
void display_delay(unsigned char delay_counts)
{
    unsigned char i;
    int  j;

    for (i=0; i<delay_counts; i++)
        for (j=0; j<5000; j++)
            ;
}
```

Figure 3-2 Original Software Listing for display_delay().

Reader Instructions: Is Ravi right? Will changing the input argument to this function fix the problem? Does this give you more hints regarding the big problem?

The function was short.

"It's not absolute time; just a loop within a loop." Ravi turned to Oscar. "The loop just counts up, doing nothing. I have seen these before; it just uses up processor cycles and creates a time delay." His smile spread as he continued, "Since we upgraded to faster hardware, this loop takes a shorter amount of time to run, so the animal faces appear faster! That's it - we make the number bigger!"

"And that, my friend, is the source of your bug." Oscar stood up and motioned Ravi to exchange seats with him. "Now, normally I would let you finish this up by yourself, but we have a hard deadline, so let's do this together. Let me drive."

Oscar pulled the keyboard to him and began typing. Playing around in the code was a passion, he admitted to himself, and solving this problem had already captured him.

"I suspect your first action would be to just choose a larger number and try it out. However, trial-and-error on this one will take too long. There are two reasons why. Well, really one *main* reason - the design of the main loop sucks. I would like to rewrite it to display an animal and then wait for the animal sound to finish playing before displaying the next one. Alternating display and sound function calls. But I checked how the sound engine works and we don't want to touch it - just let them queue all the sounds at once."

He finished launching the debugger and connected the device. "Since all the sounds are queued together, we have no idea in time when they finish. Trying to tune in the right number by running the program over and over is a crappy way to do it. So we're going to characterize the delay and use some simple math."

Ravi interrupted him, "I know - you're going to measure how long the sounds are, and then measure how long it takes the loop to finish."

"Pretty close. We need to measure the execution time for both the delay function and the function that displays the graphic. The sum of those two execution times should be the same as the sound bite so sight and sound occur simultaneously. We'll try to fix the problem by making the delay function delay a little while longer."

"How do we measure the execution times?"

"With the debugger, there are several ways to do it. Watch me. Since we have no crazy multitasking going on, I can use nonbreaking breakpoints before and after the function we are interested in." Oscar paused to gauge his reaction to the apparent oxymoron, but Ravi just waited.

"Anyway, then the debugger won't stop execution and we will get valid time-stamps for each breakpoint. We subtract successive timestamps and then we know how long it takes each function to run."

Ravi interrupted him and pointed to the listing. "But that doesn't tell us what to change the number to."

"Hold on - here are the values." Oscar scrawled numbers on a scrap of paper and then headed to a whiteboard (shown in Figure 3-3). "Let's work this out one step at a time. First, the debugger tells us that the **display_picture** function takes 102 milliseconds and the **display_delay** function takes 40 milliseconds. So the total display-and-delay execution time is 142 milliseconds for each animal."

> *Known Information - Givens*
> Duration of each sound   = 1.5 sec
>
> *Measured execution times*
> display_picture ()   = 102 msec
> display_delay ()   =   40 msec
> total time   = 142 msec

Figure 3-3 Whiteboard Notes for Known Information.

Reader Instructions: Oscar's whiteboard notes are shown in Figure 3-3. Based on these measured values, can Eduardo's analysis be correct? Also, what can you derive about the delay function? You now have enough information to propose an acceptable solution - list your proposed changes before continuing.

Ravi consulted his computer and relayed, "If I multiply 142 milliseconds by 12 animals, I get 1.7 seconds to display the entire opening screen. That matches what Eddie reported in the CR - about 2 seconds until the program ends prematurely."

Oscar nodded. "So far, so good. Now we calculate the new time delay we need to make the sound and pictures match up. If each sound bite is 1.5 seconds and the display routine currently takes 102 milliseconds, the delay function must be changed to waste 1.398 seconds. Now how do you compute the new reload value to do this?"

"I would figure out what the value 198 means in time. We need to know how much time each count really takes before we find a new reload value. And since the code is a nested loop, the function actually loops 198 * 5000 times."

Oscar grabbed a calculator. "The current time delay for 198 is 40 milliseconds. Dividing 198 * 5000 into 40 milliseconds gives a time delay of 0.04 microseconds for each count. Pretty darn small."

He continued to write on the whiteboard, thoughts racing ahead to the magnitude of the new reload. "Finally, we compute the new loop counts needed to waste 1.398 seconds. Since each count is 0.04 microseconds, that comes out to be . . ." Punching again at the calculator he concluded, "about 35-million counts."

After working his own calculations, Ravi added, "That means that the reload value for the outer loop now becomes 6990. That would be the 35-million number divided by the inner loop counter of 5000, right?" (Their calculations are shown in Figure 3-4.)

"Bingo. Now, that number may not be exact because we made some estimates, but it will be pretty close."

"That's it? That was a lot easier than plugging in numbers starting from 198." Ravi headed back to his bench. "I will go make the change now and tweak it in."

Oscar was standing to follow him when his cell phone beeped with a text message from Randy asking for the final code. They were nearly done, and he glanced at his watch to gauge their progress before calling out to Ravi while answering Randy's message.

"Again, Ravi, think before you code. If you just change that number, will the code work? Quick hint - no."

Ravi turned back. "What do you mean? We just worked it out - that value will delay the right amount of time."

"In an ideal world, yes. What will the **display_delay** function do with a number like 6990?" He tapped his fingers on the edge of the bench, and then pointed to the function declaration. "The variable is an unsigned char. What will happen?"

<u>1) New Time Delay Needed:</u>

$$\text{delay} = \text{sound time - display time}$$
$$= 1.5 \text{ sec} - 0.102 \text{ sec}$$
$$= 1.398 \text{ sec}$$

<u>2) Find Time Elapsed for each Loop Count</u>

$$1 \text{ count time} = \frac{\text{total delay time}}{\text{total loop counts}}$$

$$1 \text{ count time} = \frac{40 \text{ msec}}{198 * 5000 \text{ counts}}$$

$$= 0.04 \frac{\text{usec}}{\text{count}}$$

<u>3) New Counts Needed:</u>

$$\text{New time delay} = 1.398 \text{ sec} = \text{new counts} * \frac{1 \text{ count}}{0.04 \text{ usec}}$$

so...

$$\text{new counts} = 1.398 \text{ sec} * \frac{1 \text{ count}}{0.04 \text{ usec}}$$

$$= 34,950,000$$
$$= 6990 * 5000$$

$$\therefore \text{ New reload value} = 6990$$

Figure 3-4 Whiteboard Notes for Delay Mathematics.

"The max value is 255. Oh, I didn't think of that."

Ravi dropped his chin in one upturned palm, staring at the code with a confused look on his face.

"Change the variable to an int and you'll be okay for delays up to 32,767."

As Ravi bent to work, Oscar tapped out a response to Randy, but before he could finish the text message, Ravi groaned and waved him over.

"I found other references to this delay function." He leaned aside for Oscar to look. "It's buried in the sound-driver code, even though that code has nothing to do with the display. It looks like they're also using it for a generic delay." Ravi looked up at him and asked, "If we change the delay code, what's going to happen to the rest of the system?"

Ravi's words hit him like a brick, and Oscar felt his face get warm. He glanced back and forth between the function calls and realized Ravi was right.

"Crap." He dropped down on a stool, his mind racing between the deadline and the new information. The software was more unstable than he thought. If they fixed the delay here and the function was called elsewhere in the code, they might break other software without realizing it.

Something nagged at him not to change the main interface, but to leave all the 198s and scale the function internally to contain the code change. To further minimize risk.

This seemed to be spiraling out of control.

Ravi looked to him nervously, clearly unable to provide a robust solution.

Exhaling forcefully with his decision, he motioned again for Ravi to let him drive. "We can't change the display calls in the main loop. We have to change the delay routine itself so it still accepts 198 but converts it to the new delay value, and hope that the scaling translates to the rest of the system as well. It won't be exact, because we don't have floating-point math, so the speed ratio will be truncated to a whole number."

The editor was open to Ravi's changes, and Oscar quickly added to the file. (See Figure 3-5.)

```
#define HARDWARE_UPGRADE_SPEED_RATIO    (6990/198)

void display_delay(unsigned char delay_counts)
{
    int  i;
    int  j;
    int  new_delay_counts;

    /* CR060805 Convert incoming delay argument to new loop counter value
       to support new (faster) hardware. Measured conversion for old reload
       of 198 becomes 6990. */

    new_delay_counts = delay_counts * HARDWARE_UPGRADE_SPEED_RATIO;

    for (i=0; i<new_delay_counts; i++)
        for (j=0; j<5000; j++)
            ;
}
```

Figure 3-5 Final Software Listing for display_delay().

"Check what I did, figure out the error and tweak in the timing. Then test it for basic functionality. I'll let Eddie know it's coming, okay?" Oscar pushed off from the bench to leave, but changed his mind and turned back.

"Add a comment block for that function to explicitly state that it depends on the underlying hardware platform. Also put in that time-per-count conversion factor we derived and the maximum allowed input value to the function. That'll help the next person."

Ravi felt nervous walking to Oscar's cube. The day had been a rush of emotions and he felt a little worn out. The changes were done and tested. Working with Oscar had been stressful, but oddly enjoyable. And he felt he'd learned something. It would have taken him a lot longer to find the bug on his own; Oscar just seemed to know where to look. Taking a breath, he tapped the edge of the cube and stuck in his head.

"I made the changes and tuned in the reload value. It was pretty close; the timing error is less than 1%. Eduardo's rerunning the system test now." He paused, and added, "Thanks for showing me how to find that problem."

Oscar tipped back in his chair and nodded. "It was a good experience."

"I had a question, though. Why couldn't we just write a real delay function, one that measures milliseconds or something like that?"

"That's a good thought. But what we have here is something called 'legacy code.' Working code from the past that we don't fully understand, but we have to maintain. Truthfully, the way that sound system is implemented scares me and I don't want to mess with the timing in that subsystem. I don't want to risk touching the timer subsystem and interrupts just to rewrite the display delay function. With legacy code, it's better to make only small isolated changes that affect the fewest number of systems, and test each change individually." [1]

"That makes sense." Ravi crossed his arms. "I also had another question. That Pong program. You had that old program here, at work, after all this time?" It seemed insane that he had that program so close at hand, coincidentally on the exact day he needed to show it to Ravi.

"I don't throw anything out. The joys of a dirt-cheap data storage; everything's with me. All the software I have ever written."

"That's how you knew what my bug was."

"Turns out the bug in the Pong program is very similar. I generated the delays by calling a library function that was provided by the compiler vendor. The function generated delays in one-millisecond units, and the delay length was specified by an integer argument. It turned out the vendor implemented the delay function as a simple decrement loop calibrated to the standard speed of the PCs of the time. Newer systems are much faster, screaming along at 3+ GHz, so the 'standard millisecond' using that delay loop is now *substantially* shorter. "

Ravi blurted, "And that's why the little ball is screaming across the screen!" Oscar shared in his humor.

"And just one more question." Ravi couldn't believe he was about to make the suggestion. "Li Mei heard that sometimes people go to Molly's after work and I told her that we should go. I thought maybe all of us. I mean, if you want."

Oscar continued to rock back in his chair but he didn't answer right away. Ravi bit his lip. Just because Oscar helped him with a debugging problem didn't mean he wanted to hang out with him. He was the manager. But before Ravi could retract the offer, Oscar nodded at him.

"Actually, that is an excellent idea. All three of you have finished projects in the last couple of weeks and Li Mei is a new member. It's high time we all go out and celebrate."

"A toast." Oscar raised his glass, and Josie, Ravi, and Li Mei followed suit. "We're here to celebrate some tough project deadlines and a new employee. So here's to dead bugs and clean compiles."

Cheers from around the table mixed with the background noise of the pub. Oscar looked around at everyone laughing and helping themselves to plates of nachos, potato skins, wings, and dip.

His team.

It would be difficult to do this right. To balance needs. He'd practically taken over Ravi's bug fix to get it done in time. But the experience of the last few weeks seemed to click something in his head. He saw Josie raise her glass.

"Another toast." Josie looked straight at him as she spoke. "This one's in honor of Oscar: a quote befitting our daily activities. 'Computers allow you to make more mistakes faster than any other invention in human history with the possible exception of handguns and tequila.' Here's to his teaching us to solve them." [2]

"Oh, that's a good one!" Li Mei took another drink of stout and grimaced. "But I think I do not like black beer." She turned to Ravi. "Did you get your code review approved too?"

"Well, I guess so." Ravi looked uncomfortably around the table.

"That's good. My Meter Magic is completely done now, even the new functions. What was wrong with your program?"

Li Mei's forwardness surprised Oscar, and he wondered how honestly Ravi would respond.

"Well, I had to port software to a new set of hardware, but it turned out to be more than just remapping hardware lines." He described the animal preserve product and explained how the old program had implemented a time delay. "Even the variable initializations didn't port over correctly. I can't believe how much could break just moving to different hardware."

"Another round, folks?" The waitress dipped her head in, patting her sleeve dry with a towel. "I am sorry," she added with a laugh, "Johnny tapped another keg and spilled it everywhere. It's fresh if you want it, Oscar."

Josie smiled and elbowed him. "First-name basis, huh?"

"Thanks, Maria, I'll have another." He drained the rest of his glass. "Say, you're a dishy one. You want to come home with me tonight?"

"Sure, I'm off at 8." Maria mopped up a spot on the table and turned to wink at him as she left with the empty glasses.

The table was suddenly silent.

Oscar straightened sharply and announced with a finger in the air. "This reminds me of another quote. 'Programming is like sex: one mistake and you have to support it for the rest of your life.'" [3]

"Here you go, Oscar, fresh as it gets." Maria carefully set the full glass in front of him.

"Thanks, Maria, you're the best sister-in-law in the world. I'll call Toni and tell her you're coming over."

"I can't believe you did that! I knew you'd never cheat on your wife." Josie smacked him in the arm as he tried to open his cell phone, and he scrambled to catch it before it hit the floor.

"Oh, I'm sorry!" Josie bent to assist, but he waved her off.

"'Tis but a scratch. Just a flesh wound.'" [4]

"Oh, no. I get it," Maria groaned. "It's not me you want to see. You just want me to bring over more Monty Python videos."

Things Ravi Learned about Debugging Today

Specific Symptoms and Bugs

- *Initialize all variables. Don't assume the compiler will do it for you.*
- *Document any underlying hardware assumptions.*

General Guidelines

- *Debugging tools allow simple timing characterizations without stopping the program execution.*
- *Randomly making changes is not the sign of a good debugger.*
- *Bug report descriptions can be misleading.*

Chapter Summary: The Case of the Abbreviated Program
(Difficulty Level: Easier)

Upgrading an existing embedded information device for wildlife preserves to newer hardware causes unexpected problems when previously functional software breaks. Porting is more than obvious changes (I/O port lines, buttons, LCD display) and subtle interactions involving timing are caused by old hardware-dependent code. A hard deadline forces a trade-off on possible solutions to reduce risk and System Test burden.

The Problem Symptom(s):
- After the animal device powered up, it displayed a ring of animal faces and then quickly powered down.
- Appearance of animal faces did not coincide with audio playback of each animal voice.

Targeted Search:
- Methodically verified previous hardware port changes using a pencil, old and new schematics, and old and new software.
- Brainstormed the cause of immediate halt and display-sound mismatch using information from the System Test report.

The Smoking Gun:
The delay routine used to synchronize presentation of animal faces and voices used hard-coded loop counters to create a time delay.

The Bugs:
Two bugs directly related to porting software were found:
- The variable that ends the program was "accidentally" initialized to the ENUM value for QUIT, causing a premature end to the program. (After a compiler organizes the application object files into text (code), data (initialized variables), and bss (uninitialized variables) sections, the linker creates an executable that causes these sections to be loaded into the appropriate memory locations (text first, then data and bss). These locations are system dependent, but generally include ROM, disk, FLASH, and RAM. Vendor-supplied startup code normally handles initializing data and clearing the bss section before jumping to main(), although old startup code may not do this and variables may end up with weird values (for instance, a pattern left over from a power-up RAM test). As a rule, operations on uninitialized variables are undefined. Initialize a variable or assume the value is garbage.)

- A for-loop was used to generate a time delay. (On slower hardware, these do-nothing commands take noticeable time to execute. This was a common time-delay method. However, compilers that optimize programs for speed look for do-nothing code like this and remove it. Therefore, this calibrated delay loop would execute in zero time. This compiler optimization can be defeated by including a simple global function call - that doesn't have to do anything - in the inner loop.)

The Debugging Method Used:

- Characterizing the device using *black box debugging* without looking at the software. Two major symptoms were identified that directly correspond to the two critical bugs later found.
- Using *white box debugging* methods including code reading to understand the main loop.
- Using debugger breakpoints to measure execution time of different functions in order to compute new loop indices.

The Fix:

- Initialized the **selected_animal** variable.
- Changed the delay function internally to accept the old input arguments, which were then converted locally to new loop counter values to mimic the old observed time delays. This reduced the risk of introducing other problems, and reduced the number of required code changes.

Verifying the Fix:

Used nonbreaking breakpoints to measure the new delay time, and to verify that animal faces and sounds occurred simultaneously.

Lessons Learned:

- Make small, planned changes to legacy code, testing each in isolation to ensure nothing else is affected.
- Don't assume anything will work correctly when porting code to new hardware or a new compiler. Methodically locate and test all obvious and subtle interfaces between hardware and software.
- Expand your focus beyond the part of the system you touched.

Code Review:

This software segment is reasonably self-documenting, although it is not apparent that the animal pictures and sounds should occur simultaneously. Explicitly triggering both to occur simultaneously is desirable. Hard-coded delays of 198 aren't in real units. Consider a CR to evaluate rewriting the function to accept time in real units. This would require changing all software that uses the function, so the benefit should be weighed against the risk.

What Caused the Sea Launch Failure? The Sea Launch mission was doomed by a change to one conditional statement in the code that did not initialize a variable, leaving a helium valve open prior to lift off. An excessive amount of gas leaked away, leaving nothing to pressurize the second-stage fuel tanks. The rocket could not reach orbital velocity and the flight was terminated.

The million tests automatically conducted in the hours prior to launch missed the one-line code error.

The lesson is that software is very complex and tests don't check everything. However, code coverage mandated by the DO-178B Level A software standard would have picked up the problem. Though this is an expensive standard to support, it's a lot cheaper than the $100-million lost payload. [5-6]

References

[1] Feathers, M. (2002), "Working Effectively With Legacy Code," Object Mentor, Inc. Accessed from *www.objectmentor.com/resources/articles/ WorkingEffectivelyWithLegacyCode.pdf*.

[2] Wikiquote, "Quotes about Computers and Computer Technology," Mitch Ratcliffe. Accessed online September 9, 2006 from *http://en.wikiquote.org/ wiki/Computers*.

[3] Quote Garden, "Quotations about Computer Programming," Michael Sinz. Accessed online September 9, 2006 from *http://www.quotegarden.com/ programming.html*.

[4] Gilliam, T. (Director) and Jones, R. (Director), (1975), *Monty Python and the Holy Grail* [motion picture]. United Kingdom: Mark Forstater and Michael White.

[5] Boeing (July 1, 2000), *Sea Launch - Summary of Investigation and Return- to-Flight Preparations - July 2000*, news release. Accessed from *http://www.boeing.com/news/releases/2000/news_release_000714t.html*.

[6] Sea Launch (2000), Past Launches: ICO-F1. Accessed from *http://www.sea-launch.com/past_icof1.html*.

Additional Reading

Allen, M. (2002), "Bug Tracking Basics: A beginner's guide to reporting and tracking defects." *STQE* magazine, Vol. 4, Issue 3, pp. 20-24. Accessed online May 18, 2006 from *http://www.stickyminds.com/ sitewide.asp?Function=edetail&ObjectType=ART&ObjectId=5898.*

Ganssle, J. (2004), *The Firmware Handbook*, Burlington, MA: Elsevier (Newnes imprint).

Chapter

The Case of Thermal Runaway: Rare Transient Bugs Are Still Bugs

[*Author's Note*: A different version of this mystery first appeared in *Embedded Systems Design* magazine, 2004 [1]].

As Josie walked back from the cafeteria with Li Mei and Ravi, she reminisced about the team's first night out a month earlier. Oscar's unexpected invitation that afternoon had been a pleasant surprise. They'd been at Molly's Irish Pub & Grill for nearly three hours, talking and getting to know one another better. She had been pleased to see Li Mei and Ravi engaged in conversation; they were both the newest members of the company. She hoped the team would pull together and become one of those really fun groups to be in - a group that gelled and excelled.

Molly's had broken the ice, and over the past month they ate lunch together frequently. She was finding herself in a different team role than she had imagined. Ravi and Li Mei had both begun to ask her questions that carried a subtle respect. It was a new and pleasant thing, Josie mused, this implicit trust. Li Mei even more so than Ravi, who seemed to be more reserved. The team was taking shape, and it made her look forward to coming to work in the morning.

When they reached Ravi's cube, he peeled off with a nod and Li Mei followed Josie into her cube. Li Mei was engrossed in a monologue about her first attempt at swimming the previous weekend. Josie smiled at her enthusiasm; Li Mei seemed willing to try anything once. She happily relayed her near-drowning at the Jersey shore. Josie logged into her computer as Li Mei continued to chatter about the crowded Saturday at Jenkinson's, visiting penguins at the aquarium and eating pink-and-blue cotton candy.

On spotting a high-priority email from Oscar, Josie held up a hand. Li Mei fell silent, and at Josie's invitation began reading over her shoulder.

-------- Original Message --------

Subject: Industrial Enclosures Corp. Malfunction

Priority: High

From: Oscar Shelley <oshelley@hudsontechnologies.com>

To: Josie O'Neil <joneil@hudsontechnologies.com>

CC: Li Mei Cheng <lcheng@hudsontechnologies.com>

Josie,

We've got a potentially expensive problem. Sophie - production manager over at Industrial Enclosures - told me last week a small oven on a manufacturing line malfunctioned. Rather than heating components to a predefined temperature, it didn't shut off and damaged them. They shut the line down for the rest of the afternoon, and it started working the next morning, so they didn't call us. Unfortunately, it failed again yesterday afternoon. She says that they have a critical delivery deadline that cannot be missed. Actually, what she said was that it **Cannot Be Missed**.

Take Li Mei - go have a mentoring moment. Get it fixed. Let me know.

Oscar

"Hmm." Josie scanned the email but couldn't recall anything about this machine. She turned to Li Mei. "Looks like we are going on the road for the first time together. But, unfortunately," she sighed, "we're not going to the shore."

"Rats. Maybe we can go as a team sometime." Li Mei asked, "Did you work on this oven project before?"

Josie grinned at the mental image of staid Oscar swimming through the surf, but kept the thought to herself. "No, this is a first for me." She noticed that Li Mei was sitting straighter in her seat, and she suspected Li Mei was a little nervous about her first site visit. Walking into a customer's site with very little knowledge about a problem could be nerve-wracking and was something that still made her nervous.

She sent Oscar's email to the printer and turned back. "Hey, I know site visits are stressful, but try not to be nervous. Everything we do should just be methodical and logical, and we'll find the problem, okay?" Li Mei nodded imperceptibly and she continued. "First, we make an action plan. When we get there, we will talk to Sophie and learn everything we can about the machine and the problem. Next, I generally talk to anyone who actually saw the system fail."

"Wait a second," Li Mei interrupted, and then ducked out of the cube, calling behind her, "Let me get my notebook." Moments later, she reappeared and began writing:

Action Plan for Solving Problems
1) Interview problem reporter
2) Interview anyone who saw system fail

Josie nodded and continued. "After I have learned everything I can by talking to people, I want to see the system fail because people's memory and descriptions of the problem can be bad or incomplete. Sometimes we can make the system fail pretty easily, but other times the failure is intermittent and it can be a pain to reproduce it." She motioned to Li Mei's pad. "You could call this next step 'Observing the system behavior.' You want to see normal and abnormal operation if possible, so you understand how the system is supposed to run under normal circumstances."

Balancing backward in her chair as Li Mei jotted, she reflected for a moment and laughed before continuing.

"You have to be careful about deciding what's *normal* behavior; some machines shudder and make loud noises, and you think that the building is going to come down, and it turns out that *is* normal for that machine."

"Really?" Li Mei underlined the word "normal" on her pad and then looked up. "Do they let you run the machines? What if the machine is part of a big assembly line? Can we stop the line? They will listen to us?"

"It depends. You'd *think* they would let us do whatever we needed to fix the problem, but sometimes they won't shut down the line if the machine is working well enough to keep up production. It can be a real pain - you're standing there looking at a machine that the customer called up ranting about, and then they won't let you get close enough to see what's wrong." She leaned forward to tip her chair level again and lowered her voice. "Some customers are real jerks and you have to deal with it. You just can't lose your cool."

Li Mei giggled and started to put pen to paper, but Josie stopped her. "Don't write that! Just remember that rule in your head."

"Okay, I will make a special rule: 'Interview customer and keep cool, even if the customer is not nice.'"

"That's perfect." Josie smiled at her. She didn't add that Li Mei's special rule would probably rank high on an unwritten list that would most certainly grow after she'd made a few more customer visits.

———————————————————

By the time they arrived at Industrial Enclosures it was late afternoon and the shift had ended. Sophie corralled them into the production area, giving a terse overview of the company as she led them serpentine through the machinery and assembly lines. Li Mei clutched her notepad and glanced everywhere while straining to hear Sophie explain that Industrial Enclosures manufactured custom enclosures for different types of industrial equipment, producing more than 100 different product lines each year.

Li Mei was amazed at the amount of equipment and raw materials stacked about, especially in such a small shop. As they walked, the acrid smell of hot plastic became unpleasant.

Almost as an afterthought, Sophie glanced back to make sure Li Mei was following before declaring, "The manufacturing line that failed was used to assemble enclosures with airtight compartments. We are using a special material to construct these compartments. The manufacturer of WGM-17 has specified its temperature characteristics - here are the specs." She handed Josie a sheet of paper (Figure 4-1) and continued matter-of-factly.

"The operating range is between 117°F and 165°F, and the recommended nominal operating temperature is 126°F. The manufacturer told us all bets are off if the material is heated beyond 189°F. Structural characteristics are out the window. Got it?"

WGM-17 Thermal Properties and Operating Limits	
117°F	Minimum working temperature
126°F	Nominal working temperature
165°F	Maximum working temperature
189°F	Material loses structural integrity

Figure 4-1 Temperature Characteristics for WGM Component Material.

They reached a small conveyer line near the wall, and stopped. Since work had finished for the day, the line was shut down and nearly everyone had gone home. Sophie gestured to the metal box that enclosed a portion of the line.

"We use this small oven that was specially designed to heat the components to 126°F so that each component makes a proper seal with the underlying structures. However, the oven didn't shut off when it reached 126°F and it heated the components to the point that they were unusable."

Sophie turned to her silent audience and focused on Li Mei. "I believe you have already detected the evidence of compromised material, perhaps?" Li Mei stared back, uncomprehending. After a short pause, Sophie pointed to her nose.

Then Li Mei quickly nodded. "That bad smell is from the oven? Yes, it does smell bad."

"Luckily it isn't toxic when it overheats. It just relaxes and molds to whatever it touches. But I am left with this psychotic oven and a truck coming the day after tomorrow to pick up a shipment of 200 finished enclosures."

Sophie crossed her arms and leaned against the assembly line to face them both.

"That's where you two come in."

Li Mei stared at her frankly, wondering how a pretty, middle-aged woman had ended up as plant manager of a loud, smelly factory. Sophie seemed all business, and Li Mei felt more and more out of her element as Sophie's large green eyes drilled into her. So far, Sophie hadn't smiled and Li Mei found herself unable to think of anything useful to say.

Luckily, Josie had already started to ask questions.

"Both times the oven failed, did it fail the same way?"

Sophie tipped her head forward to think before replying. "I believe so. Only BJ saw the first failure; he's the operator for this line and he's gone for the day. We run a 7 to 3:30 shift. You will be back then?"

Josie glanced at Li Mei before answering. "We will be here at 7 a.m. to see BJ. Can you tell us anything about the design of this oven, like about the hardware or software?"

"It's microprocessor-controlled with software customized to the material requirements. The engineer who designed the system is no longer with us."

It was then that Li Mei realized that Hudson Technologies hadn't designed the oven, but was probably just a consultant on the project. That's why Oscar couldn't provide them any documentation. Deep in thought, she almost missed Josie asking the very question she had in her mind.

"Did the engineer leave any software code listings or documentation for this system?"

Sophie nodded as she pushed off the assembly line. "My assistant will have it for you in the morning. I'm aware this is tight, but I'm putting an incredible amount of faith in the two of you to identify the cause of this failure."

As they drove back to Hudson Technologies, Li Mei felt like her head was full and empty at the same time. They had learned nothing and the visit seemed to be a waste of time, and she told Josie so.

"Well, not completely a waste," Josie said. "We got to see the machine and learn some background information, and we got the specs on the material. Start playing detective. It's a thermal control system with a specific temperature band. What do we know?"

Li Mei wasn't sure what Josie was getting at, so she kept quiet.

"Think. We have a thermal control system. It's broken. But only sometimes. What are some common failure points for temperature-controlled systems?" Josie tromped the gas to make it through a yellow light before she continued.

"I'll give you a hint. The oven's gotta have a heating element. Duh. It could have failed, or gotten dislodged or been damaged. That could cause the thermal system to malfunction." She paused, and then prompted, "Now you think of one."

Li Mei stared out the window and tried to imagine what else could go wrong, and then realized an easy one. "I know - if it controls temperature, then a temperature sensor must be there. It could have the same problem - be broken or moved around." She looked to Josie for confirmation and was rewarded with a nod as Josie continued to work her way through the start of rush-hour traffic.

"You are absolutely correct. Good one. Now see how logical this can be if you just go one step at a time?" Josie glanced at her quickly before looking back to the road.

"I'll go again. Sophie said there's a microprocessor in there, which means it's probably running a temperature-control algorithm. Something that senses the current temperature and tells the heater to turn on or off to keep the oven at the right temperature. Something could be wrong in that software."

Li Mei shot back, "But if there was a bug, how could the oven work at all? Wouldn't it fail all the time?"

"That's a good point. It looks like we have an intermittent problem, one that occurs every now and then. Something that *seems* like hardware is causing it. An obvious software bug would happen more regularly, but a subtle bug can display intermittent behavior. Now your turn."

This time, Li Mei was stumped. It seemed like a lot of work to think of all these possible problems when tomorrow they could go in and just look at the software. She watched the landscape go by; fast food, gas, minimall. Josie signaled to turn and they cruised past Molly's and into the business park where Hudson Technologies was located. Pulling into a parking lot, Josie broke the silence.

"Another obvious failure point. Say the temperature-control algorithm decides that the oven is too cold, and wants to jack the heat up. What happens?"

Li Mei leaned her head back in the seat and imagined the software code.

"Oh, I know - the hardware that controls the heater. A heater draws a lot of current, so the microprocessor has an output port that controls a relay or something. The port or the relay could be bad."

Josie shifted the car into park and pronounced, "Bingo - she's a winner. Meet you here tomorrow morning around 6:30 a.m."

After picking up an extra-large cup of dark roast coffee on the way to work, Josie pulled into the lot to find Li Mei sitting on the hood of her old red Honda, sipping from a plastic mug. She looked so tiny, cross-legged and hunched over her tea, straight black hair tucked behind her ears and flowing halfway down her back. Josie wasn't surprised to see that Li Mei had beaten her to work; she seemed to have a great work ethic for someone so young. She stopped in the nearly empty lot and waited for Li Mei to join her.

"Morning, Li Mei. Ready for action?"

"I thought that if the temperature-sensor hardware, like A-to-D converter, is bad, this could cause the problem. I am ready to find this heater problem and fix it up!"

"Hey, Li Mei, I'm doing good today, too."

Li Mei stopped short and looked slightly crushed, but Josie laughed and pulled back out of the lot. "Good attitude - let's go fix it up."

At Industrial Enclosures, they found a tall guy leaning over the oven assembly line, fiddling with something out of view. When he saw them, he stood up and introduced himself.

"Hi there, I'm BJ. Sophie told me some folks would be out to take a look at this oven, gone all haywire and messing up parts." He offered his hand and introductions passed all around. Josie brought him up to speed on their conversation with Sophie, and their initial thoughts about what might be wrong.

Then she asked, "Would you describe what happened when the machine failed?"

"Sure. Can do." BJ began to explain, "The oven normally takes a couple of minutes to slowly heat the WGM-17 to the proper temperature, and when the component is hot enough, the oven automatically turns off, and the component moves out of the heating area. I noticed that heating started to take a lot longer, almost five minutes." He patted the top of the oven, "When I looked inside, the oven was very hot so I hit the emergency shutoff. When I picked up the component with tongs, the tongs made imprints in the material. It's not supposed to do that."

His expression was almost comical: the idea of his machine misbehaving! "It looks like the oven didn't turn off when it was supposed to. I have been running this machine for about nine months so I know what it's supposed to do."

He reached under the line and retrieved a small object, handing it to Josie. "This is one of the overheated components from the other day."

She turned it in her hands. It had definite deformations when she compared it to a properly heated component he proffered with his other hand. Looking up she asked, "Does this happen all the time, or just one unit every now and then? Is it reproducible?"

"Well, for about an hour on Tuesday afternoon the oven continued to fail, and we shut the line down for the day. Wednesday morning it started working again and worked all day, no problem."

She thought about his words and refined her questions. "BJ, you said that the oven 'continued to fail.' Does this mean it overheated the component every single time or just every now and then?"

"Oh, no doubt there! After it fails once, it fails every time." He leaned in conspiratorially. "Of course, the components aren't cheap, so that Sophie, she shut down the line right quick." BJ twirled another of the melted components between two fingers, clearly enjoying the process.

"Okay." Josie thought about the brainstorming that she had already done with Li Mei. "We had some thoughts about the system, and what might cause a thermal failure. Is it possible the heating element failed or moved inside the oven?"

"No, the heater's mounted in there pretty rugged."

"What about the temperature sensor? They're generally more fragile. Could it have moved too far away and measured a lower temperature? That could cause the heater to overheat."

BJ nodded. "I was thinking that too, but I replaced it and the oven still overheated. It has a permanent mounting."

"Has the oven ever overheated before when it was first turned on?"

"Well, once we had a problem calibrating the temperature sensor, and the oven did overheat that time. I checked the oven temperature this morning with a different temperature probe, and it's within one degree of normal. It's good to go."

Josie consulted her notepad, and remembered her next line of questions. "Is there a temperature set-point knob or any other adjustments on the machine that could have been moved? Any recent maintenance on the machine?"

"No maintenance lately. Here's the temperature set point; it's dialed to 126°F."

"BJ, I'm not sure what hardware controls the heater but probably something like a solid-state relay. Could that be bad?"

"Well, it's possible, but the oven appears to be working now. If the problem happens again we can check it."

"Was any new software installed on the machine?"

"Nope."

"Did someone else operate the machine?"

"Absolutely not."

She sighed. No hits at all. Li Mei hadn't moved during the exchange, but stood furiously writing notes on her pad.

As Josie looked over her list and pondered what else to ask, BJ moved away and started up the line. They watched one component at a time advance down the line and enter the oven. After a short time, the oven clicked off and the component was ejected from the other side. BJ picked it up with a pair of tongs and inserted it into a larger assembly, making sure it was properly placed. It took him almost a minute to perform the operation. Clearly, BJ was in work mode now, and it seemed from his perspective that everything was working fine.

Josie rocked from side to side and mentally admonished herself for wearing boots; they might be standing here all day waiting for that damned oven to fail again. Well, she admitted, Oscar wanted a mentoring moment, so now was as good a time as any.

"Li Mei. So we've blown through the first couple steps of our action plan with no success. Nothing obvious in hardware and we don't have the software listings yet. So now we think about what type of failure we have, and the possible root causes for that type of failure." She thought for a moment, and continued.

"Some events are one-time repeatable events. These symptoms occur once, but have a pattern to their occurrence. They might occur only at power-up or the first time through a function or feature. Or, a function may work correctly the first time but fails all subsequent times."

Li Mei nodded for her to continue.

"With periodic events, symptoms occur several times with a regular pattern. That makes them easier to trap in software.

"The trickiest can be sporadic events. These may happen once in a hundred tries or so randomly that it's hard to relate the occurrence to software."

She directed a query at Li Mei. "How would you classify the oven symptoms?"

Li Mei was immediately ready and blurted her answer in a single breath. "If I have to choose between one-time repeatable, periodic, or sporadic event, I am choosing 'sporadic event' because the oven stopped working and then started working again with no reason why."

Josie nodded. "Unfortunately, I must agree with you. Sporadic events are the hardest to find."

"I know these are the hardest." Li Mei smiled. "You think I am not as smart as you, but I will be very smart soon!"

"Hey - fine by me. You go fix this and deal with Sophie." Josie stretched to quell the ache that was starting in her back. "So then you tell ME how to debug a sporadic event!"

"No, that is okay," Li Mei conceded, while retaining the gleam in her eye. "You may continue, and I will carefully document your wisdom."

"Great." Josie sighed in mock defeat. "I spill my wisdom to the unappreciative masses." She was starting to like Li Mei's sense of humor and found the banter relieved the stress of uncertainty.

Resuming her train of thought, she continued, "The bug is sporadic, although once it fails it continues to fail, so don't forget that aspect. We design systems to behave in a repeatable manner, and when they don't, some of our assumptions are generally incorrect.

"There are several root causes." As she recited, she ticked off each item on her fingers.

"Violation of boundary conditions in the software; unexpected input or output conditions - these can be in the software, the hardware, or the material; unhandled error conditions or faulty logic, logic based on time-of-day . . . " She paused and then added, "Memory corruption, timing, and performance issues are really tough causes to nail down. And then there are intermittent electrical or mechanical connections, or impending component failure."

Li Mei started to glaze over at the thought of all the possible problems they faced. But until they got a software listing or the oven failed again, they had no other option but to brainstorm ideas.

"Again, let's take it logically. We need to narrow down these possible root causes based on what we know. Can you think of anything special about the failure that might have caused the problem?"

Li Mei chewed on the end of her pencil and offered, "Both failures happened in the afternoon. Does the oven have a real-time clock?" With a needling grin, she added, "Or, does the oven run on your favorite operating system, requiring BJ to power-cycle the machine many times for no reason?"

Josie burst out laughing. "Good thoughts. I didn't think about when the failures occurred. We'll ask BJ if he power-cycled. I was also wondering myself if the components were all made of the same material; maybe one lot from the vendor has the wrong thermal characteristics."

She walked over to BJ to ask, but returned a few moments later shaking her head.

"No luck. It still failed after he power-cycled, and all the components, both good and bad, were from the same lot."

"Maybe Sophie has some good news." Li Mei pointed behind her. "Here she comes."

Sophie greeted them both and pulled out a sheath of paper. "Here is a print-out of the software. I can set you up in a conference room with a table, and have someone get you if the oven fails again. No use standing here watching BJ work and chatting, I suppose, without making some progress yourselves."

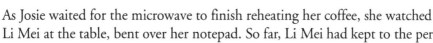

As Josie waited for the microwave to finish reheating her coffee, she watched Li Mei at the table, bent over her notepad. So far, Li Mei had kept to the periphery. Josie's own first forays into the field had been stressful, but she was getting used to it. Besides, she enjoyed meeting new people and seeing how their machines were actually used. Even though they tried to simulate in the lab how products were used, a customer always seemed to find some new and bizarre way to provoke an error condition or strange behavior.

Getting a new product ready for the market required that it be abused, on purpose, which seemed to run against a developer's natural tendency to baby it and protect it. She recalled the time when one of her previous managers had purposely taken a device she had devotedly shepherded through development and testing and threw it on the ground in front of her. She had jumped in shock, thinking he was upset with the quality of her work. But he had only been trying to make a point.

She had been overprotecting the product.

And, as he had so bluntly informed her, it would be extremely difficult to ship her along with every device. The admonition had stung; the memory was a strong one.

The microwave beeped, pulling her out of her reverie. She joined Li Mei at the table, thinking about how to present the next action step. Randomly digging through listings was inefficient, frustrating, and sometimes downright impossible, depending on the quality of the software.

"So we have this stack of software," she said, patting the listing beside her, "and we need to start understanding this system. We don't have time to get lost in the details; we need to target our search for things that look relevant to temperature control. You take part of the listing, and we'll each look for keywords like 'tempera-ture,' 'oven,' 'A2D' . . . stuff like that, okay?"

Li Mei nodded as she accepted part of the list. Josie flipped through the other half and found the code was written mostly in C with some assembly code. "It looks pretty straightforward - this software is running on an 8-bit microprocessor, if the comments are correct, and it has one interrupt service routine." She glanced at the header files and found references to simple floating-point math functions for small numbers. The interrupt service routine was called **periodic_timer()** and had several references to temperature, so she concentrated her search there. (Figure 4-2.)

Reader Instructions: Fragments of the software listings appear in Figure 4-2. Look at each of the routines and try to decipher what they do, assuming that the comments, function names, and variables are not misleading.

Li Mei looked up a short time later. "I found several functions that might be useful."

"Hmm?" Josie looked up.

Li Mei turned her pad around for Josie to see. "I found four functions we should look at. They sound like brainstorming we did before."

```
read_actual_temperature_A2D()
read_reference_temperature_A2D()
calculate_new_oven_ON_time()
oven_ON_time_control_routine()
```

She continued, "These read the temperature using the analog-to-digital converters on the microprocessor. And one is the control algorithm you brainstormed in the car yesterday. Maybe the last might tell the oven to turn on and off. I think we should look at these."

"Looks like a good place to start. I see calls to all of those functions inside the interrupt service routine." She showed Li Mei the **periodic_timer()** function. "It looks like everything gets controlled from the ISR. Every 10 milliseconds, it calls a function that appears to control when the oven is turned on. And every second, new temperature values are read and used to adjust the oven control."

Li Mei nodded. "That makes sense. Look at that last function - the one that controls the oven. I think that sets a percentage for how often the oven is on. But I don't understand what it's a percentage *of*." She pulled out a sheet containing the **oven_ON_time_control_routine()**.

"Oh, I have seen this type of control before. This is called pulse-width modulation, or PWM for short. What happens is that there's some predefined time interval, like one minute. The pulse can be ON anywhere from 0% to 100% of the time interval - the time interval never changes, but the width, or percentage, of the pulse does."

```
unsigned char oven_pulse_width_counter = 100 - 1;
unsigned char actual_temp_A2D_counts;
unsigned char reference_temp_A2D_counts;
unsigned char ovenPW         = 0;
unsigned char ten_msec_ctr = 0;
#define   Heater_output_pin          portb.2
#define   PWM_100_PERCENT            100
#define   NOMINAL_TEMPERATURE_PW     17
#define   GAIN                       2.7
#define   ON                         1
#define   OFF                        0

/* ------------------------------------------------------------------
   Name:        periodic_timer()
   Function:    Part of ISR. Every 1000 ms the new ovenPW is calculated.
   ------------------------------------------------------------------ */
void periodic_timer()
{
    /* ------- 10 msec ----------- */
    oven_ON_time_control_routine();

    ... (unrelated code removed for space) ...

    /* ------- 1000 msec ----------- */
    ten_msec_ctr++;
    if (ten_msec_ctr == 10)
    {
        actual_temp_A2D_counts = read_actual_temperature_A2D();
        reference_temp_A2D_counts = read_reference_temperature_A2D();
        ovenPW = calculate_new_oven_ON_time();
        ten_msec_ctr = 0;
    }
}

/* ------------------------------------------------------------------
   Name:        oven_ON_time_control_routine()
   Function:    Control ON duty cycle of the heater. Oven ON if conter < PWM
   ------------------------------------------------------------------ */
void oven_ON_time_control_routine(void)
{
    oven_pulse_width_counter++;

    /* Check if start of new PWM oven heating cycle; if so, oven ON   */
    if (oven_pulse_width_counter == PWM_100_PERCENT)
    {
        oven_pulse_width_counter = 0;
        if (ovenPW != 0)
            Heater_output_pin = ON;
    }

    /* Turn oven OFF at end of heating time  */
    if (oven_pulse_width_counter >= ovenPW)
        Heater_output_pin = OFF;
}

/* ------------------------------------------------------------------
   Name:        calculate_new_oven_ON_time()
   Function: Each second find new ovenPW based on current temperature
   ------------------------------------------------------------------ */
void calculate_new_oven_ON_time(void)
{
    signed char delta_temperature;

    /* Calculate new pulse width based on current temperature.  */
    delta_temperature = actual_temp_A2D_counts - reference_temp_A2D_counts;
    ovenPW = NOMINAL_TEMPERATURE_PW + GAIN*delta_temperature;

    /* Make sure duty cycle math has not exceeded 100 percent. */
    if (ovenPW > 100)
        ovenPW = 100;
    if (ovenPW < 0)
        ovenPW = 0;
}
```

Figure 4-2 Software Listing (external document).

Josie drew a train of pulses on her pad (Figure 4-3). "Say this is the signal that goes to the oven. The oven stays ON when the pulse is high - a longer pulse width means a longer ON time, or a hotter oven."

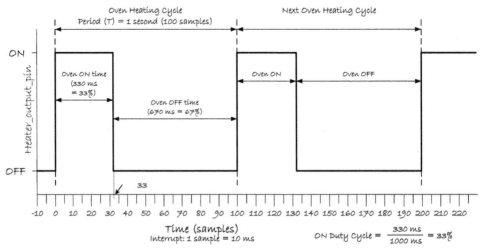

Figure 4-3 Pulse Width Modulation of the Oven ON Signal.

"But why doesn't the time interval change?" Li Mei looked up from the drawing.

"When the system is being designed, the engineers figure out a good value so the temperature can be maintained in the right range around the set-point temperature. For 126°, using a one-second interval must have been sufficient. Temperature is controlled by *how much* of that interval the heater is ON." Josie scooted her chair around next to Li Mei's.

"Let's take a look at the code together. This **oven_pulse_width_counter** variable is incremented each time the function is called, and it's reset to 0 after it reaches 100. The oven is turned on when the counter is less than **ovenPW** and is turned off all other times."

"I am confused. What is the **ovenPW** variable then?" Li Mei twirled a strand of hair.

"Well, the heating-cycle interval is one second, which is 100 function calls multiplied by 10 milliseconds per function call. That means the longer the oven heater is on during this one second, the hotter the oven can get and the more rapidly it can heat up. The variable **ovenPW** is the duty cycle of this signal and also represents the percentage of time that the oven is on during each one-second interval."

Li Mei still looked confused, so Josie added to her picture.

"Say in this picture that the **ovenPW** is equal to 33 and the oven is on 33% of the time. The **oven_pulse_width_counter** *always* counts from 0 to 100, but when it is *less* than **ovenPW**, the oven is ON. When the count is higher, the oven is off until it reaches the end of the cycle - PWM_100_PERCENT."

"Ohhhhh, I get it!" Li Mei broke into a smile. "I like that - you just tune the pulse width until the oven is the right temperature!"

Josie nodded, pleased that Li Mei was getting excited about control algorithms. Pretty cool. "Great, but before we continue, something in this function should bother you."

After a pause, Li Mei admitted, "I don't see anything wrong."

"Probably nothing is, but what would happen if **oven_pulse_width_counter** ever reached a value of 101? Can this occur, and if it did, would the oven turn off properly?"

"Oh, that would be bad. The variable would be incremented forever and never be reset to zero. That would mean the oven would turn off and never turn back on again."

"Close. Checking counters this way is dangerous because the variable is a global variable. If any other function changes this variable beyond 100, the heater wouldn't turn on again until the variable had been incremented all the way to 255 and then rolled over to 0. Better to change the logical to check if the variable is '>=' rather than just '=='." To her relief, Li Mei seemed to understand.

"Okay, now I want you to explain to me how the **ovenPW** value gets set," Josie said. She leaned back and crossed her arms, watching as Li Mei spread sheets in front of her.

"This is easy for me to explain." She pointed to the **calculate_new_oven_ON_time** function and began. "This function is basically just a linear equation with a slope and intercept. There must be two thermometers in the system; one for oven temperature and the other for reference. It takes the difference between the two, multiplies it by a gain factor, and adds an offset that must be a baseline temperature." She wrote the equation on her pad. "The **ovenPW** is 17 plus 2.7 times the difference in temperature values."

```
ovenPW = NOMINAL_TEMPERATURE_PW + GAIN * delta_temperature;
ovenPW = 17 + 2.7 × delta_temperature
```

"I can also see that it will truncate so the value doesn't go outside the range 0% to 100%," Li Mei added and looked for approval.

"I agree. And one more point. Making **ovenPW** larger will make the oven heat to a higher temperature. But the way this is coded, a larger **ovenPW** will also make it heat up *faster*. That's what happens when the oven or the component is colder. A larger **delta_temperature** value keeps the oven cycle ON more in the beginning, and when the component gets close to final temperature, the ON time is constantly decreasing each oven-heating cycle to cruise into the final temperature without over-shooting and damaging the component."

Li Mei was newly interested. "Maybe that's the problem."

"We'll check," Josie agreed. "To summarize, it looks like we have an oven that's controlled by a digital ON/OFF signal, and that the temperature is controlled using pulse-width modulation. The ON-time is computed from the actual temperature and a reference temperature, and the nominal ON pulse width is 17%, which most likely corresponds to the component material's nominal temperature of 126°F."

While they were eating lunch, Sophie marched through the conference room and placed a laptop computer on the table with the development environment. Catching them with a stern eye as she turned to leave, she said, "Now you had better get out there and figure this out. This is all I have to help you."

Josie rolled her eyes after watching the woman depart. This Sophie had been helpful but she didn't seem to trust them. She and Li Mei rose without comment to toss their garbage and head back to the factory floor.

"Okay, now we get down to business." Josie powered up the laptop and handed the software listings to Li Mei. "First, we confirm our understanding of the thermal system. I am going to set up the debugger so we can check some variables. Some logical choices would be **ovenPW, actual_temp_A2D_counts**, and **reference_temp_A2D_counts**."

"You should add **Heater_output_pin** to see if the digital heater output signal is correct."

"Yup. Good thought."

"Hey, ladies." BJ met them at the oven. "Sophie said you were ready to check the system, and I set up a separate temperature probe in the oven so we can record the component temperature directly. Will that help?" He pushed his hair back and repositioned his baseball cap, looking at them expectantly. He wasn't a typical operator; he had ownership of the line and he wanted to see things working correctly again.

She nodded and explained their plan, ducking behind the oven to plug the laptop cable into the microprocessor's debugging port. "If you holler out temperature values that would be great."

BJ started up the line and the first component entered the oven. "Right now it's 81 degrees," he called.

"Li Mei, I'll give you the variable values and you write them down after BJ gives us a temperature reading." Josie dictated numbers and Li Mei filled in the table shown in Figure 4-4.

Measured Values				Calculated values	
actual_temp_ A2D_counts	reference_temp_ A2D_counts	Heater_ output_pin	Independently measured temperature	delta_ temperature	Calculated ovenPW
[A/D counts]	[A/D counts]	1=ON, 0=OFF	[°F]	[A/D counts]	[%]
85	70	1	81	15	57
83	70	1	87	13	52
81	70	1	93	11	46
79	70	1	99	9	41
77	70	1	105	7	35
75	70	1	111	5	30
73	70	1	117	3	25
71	70	1	123	1	19
69	70	0	129	-1	17
69	70	0	130	-1	17

Figure 4-4 Data Collection Trial 1.

After a short time, the oven turned off as BJ removed the component and began to fit it into the assembly. Li Mei handed the spreadsheet to Josie and they both stared at the values.

"When the oven turned off, the temperature probe read 129°. The reference temperature is always the same." Li Mei flipped through her stack of papers and found the schematic. "Look, it isn't another temperature sensor, it's a reference voltage. The algorithm waits until the actual temperature is the same as that reference voltage."

Josie took the sheets and noted that lower A/D values were associated with higher temperatures. She nodded in agreement. "When they are the same, BJ's independent temperature reading is somewhere between 123° and 129°. That makes sense."

She turned her attention to the right side of the sheet where Li Mei had listed the corresponding delta temperature and PWM values. "In the beginning, the oven is ON 57% of the time, and then it continuously reduces its ON time as the component gets closer to 126°. So the oven is on a smaller and smaller percentage of time to prevent it from overheating, like we thought." She added, "I bet that a PWM of 17% exactly corresponds to 126° so even if the component wasn't removed right away, it wouldn't overheat."

"But something must go wrong with that logic during the failure," Li Mei added.

Josie looked back to check what BJ was doing and from the corner of her eye saw Sophie watching them. She must have been eavesdropping, Josie realized, and felt her face redden. Sophie straightened and held them both in her gaze for a moment, then nodded slightly to herself and walked away.

"Shoot, she's checking up on us." Josie sighed and muttered, "Whatever. Typical customer."

"I think she is smart, but you are right," Li Mei said quietly. "She knows I am not experienced here."

Josie cut her off, exasperated. "That's bull. Her company has a problem and we are the onsite team to solve it. Don't let a customer intimidate you - remember, they couldn't figure out the problem and it was something their engineer designed." Josie paced in a circle, thinking of her own somewhat-limited exposure to customers. Not yet 30 years old, she had already worked her way through a master's degree in engineering, with several years of experience on top of it. Her other friends from school had experienced the same attitude from many customers. It was like they were considered the source of the problem rather than the solution.

Josie stopped pacing as an idea struck her, and she headed over to BJ.

"Hey BJ, we found out that the oven turns on most when the initial temperature is the coldest. Is it possible that the material was cold, below freezing, the afternoon the oven failed? If so, that would turn on the oven 100% and maybe that causes it to overheat."

BJ chuckled, "Not likely on a summer afternoon! Sometimes, like today, we store the components in a storage area in the back. But that area has no air conditioning so they weren't frozen." Josie nodded, but after he turned his back she snagged two of the components to stick in the freezer she saw in the conference room, in case she needed them for testing later.

When she returned, Li Mei motioned her back to show her a larger table (see Figure 4-5). "Josie, I extended the table for temperature and **ovenPW** all the way to deform temperature so we can compare values if a failure occurs."

"Hey, Josie!"

At the sound of her name, Josie turned to see BJ holding tongs in one hand and waving her over with the other.

Measured Values			Calculated values		
actual_temp_ A2D_counts	reference_temp_ A2D_counts	Corresponding temperature	delta_ temperature	Calculated ovenPW	
[A/D counts]	[A/D counts]	[°F]	[A/D counts]	[%]	
111	70	3	41	100	
109	70	9	39	100	
107	70	15	37	100	
105	70	21	35	100	
103	70	27	33	100	
101	70	33	31	100	
99	70	39	29	95	Warming Needed
97	70	45	27	89	
95	70	51	25	84	
93	70	57	23	79	
91	70	63	21	73	
89	70	69	19	68	
87	70	75	17	62	
85	70	81	15	57	
83	70	87	13	52	
81	70	93	11	46	
79	70	99	9	41	
77	70	105	7	35	
75	70	111	5	30	
73	70	117	3	25	
71	70	123	1	19	Normal Operating Range
70	70	126	0	17	
68	70	132	-2	11	
66	70	138	-4	6	
64	70	144	-6	0	
62	70	150	-8	0	
60	70	156	-10	0	
58	70	162	-12	0	
57	70	165	-13	0	
56	70	168	-14	0	Too hot
54	70	174	-16	0	
52	70	180	-18	0	
50	70	186	-20	0	
49	70	189	-21	0	Deform
47	70	195	-23	0	
46	70	198	-24	0	

Figure 4-5 Valid Temperature Ranges and Corresponding ovenPW *Values.*

"The oven has failed again. It overheated a component and I've just turned it off." Josie and Li Mei jogged over, and BJ added, "Hopefully you can figure out what is wrong with it now." He lowered the tongs and placed the component out of the way on the edge of the line. It was noticeably distorted and still too hot to touch.

Josie walked over to the debugger display on the laptop computer, still connected to the control panel of the oven. Finally, an opportunity to collect data on an actual failure. BJ restarted the line, and immediately the debugger showed an

ovenPW value at 100%! Josie stared at the value, uncomprehending. BJ watched the oven and they collected data for about two minutes before leaning in to press the emergency stop button.

A small crowd gathered.

"This is exactly what happened a few minutes ago, and this is exactly what happened on Tuesday. Out of the blue the oven just goes crazy. What do you think could be wrong?"

Josie sighed and consulted Li Mei's temperature table for entries with an **ovenPW** at 100%. BJ leaned over her shoulder to look. "And those components sure aren't below freezing; we just got that new pallet of parts from the storage area and, like I told you, it doesn't have air conditioning."

"Fine - let's look at the data from this run." Looking at the debugging information (shown in Figure 4-6) she saw that the starting temperature of the components was 63 A/D counts - that's about 147°F!

Measured Values				Calculated values	
actual_temp_ A2D_counts	reference_temp_ A2D_counts	Heater_ output_pin	Independently measured temperature	delta_ temperature	Calculated ovenPW
[A/D counts]	[A/D counts]	1=ON, 0=OFF	[°F]	[A/D counts]	[%]
63	70	1	147	-7	100
61	70	1	153	-9	100
60	70	1	156	-10	100
59	70	1	161	-11	100
58	70	1	162	-12	100

Figure 4-6 Data Collection Trial 2.

"BJ, it says that this component came into the oven at 147°F. It's already in the valid temperature range, but the heater is going full blast. My table here shows that the oven should be completely off in this situation."

BJ checked the temperature of the next component in the queue and whistled, "Yup, this one's almost 150 degrees too. The readings are right."

"Why are these components so hot?" Josie was surprised to hear Li Mei join the fray. She backed up to let Li Mei continue.

"Were the components that failed on Tuesday afternoon this hot?"

"Well, we had just received a new pallet of components out of the storage area after lunch, and they were the first to fail. The storage area is next to the boiler so it gets pretty hot in that area, but it's not hot enough to damage the parts." BJ unconsciously repositioned his cap as he tried to remember. "We usually pull the pallets out before lunch so we can start the line right away after the break is over, but sometimes we get behind like today and we don't get them pulled out until much later. We might have done the same on Tuesday, but do you think that matters? The temperature of the components is still within spec."

"This is totally wrong!" Li Mei leaned over to see the debugging information displayed on the laptop and called out, "We are seeing correct values for the temperature-reading A/D counts but the oven is heating at 100%." Straightening, she added emphatically, "Totally wrong!"

Josie rocked on her heels, staring at the floor. BJ was right. Even though the components were hot, they were still within spec. The oven should not have turned on at all. Why was the oven turning on, she wondered. When she turned back to the software listing to review the calculations, she saw everyone staring at her. Even Sophie had shown up to see what she would do. Her heart started to thump as she realized time was up. She had to get the problem solved now, and she would be doing it with an audience.

Reader Instructions: You now have all the information you need to solve the mystery. Many developers have been bitten by this type of software bug. What do you think happened to cause software to induce sporadic problems? Look at the software and the last data collection run, and think.

Taking a deep breath, Josie turned to Li Mei and tried to ignore everyone else. "Let's review the facts. First, the failure happened both times when the components were already hot - around 150°. Second, when the problem happens, the oven immediately turns on to 100%. I want to calculate the **ovenPW** for an input temperature of 147°."

Li Mei grabbed her pad and wrote out the equation:

ovenPW = 17 + 2.7 × (63 - 70) = -2

She added, "This becomes -2, which is less than zero, so the **ovenPW** is reset to 0. I don't understand."

"Ahhh!" Josie smacked her forehead, and luckily stopped herself from vocalizing that the original engineer was an idiot. "I got it!"

"What? Tell me, Josie!"

"Oh, this is so common; I can't believe I didn't think of it in the first place." Josie grabbed the software listing and flipped through to the variable declarations at the top of the page.

"Look at the declaration for **ovenPW**." She pointed to the top of the page. "It's declared as an unsigned char. Do you know what that means?"

"Yes, it can't go negative. But we don't need it to be negative; just zero."

"Aha!" Josie smiled. "But when it tries to go to -2, it's actually stored as 254. And the final calculation not only rounds negative numbers up, but it rounds values over 100 down to 100%. In this case, 254 was limited down to 100, and the oven turned on full blast."

"Oh, wow!" Li Mei slowly nodded. "But then how can the '**ovenPW < 0**' ever get executed?"

"It doesn't. I didn't realize it until now, but those two lines of code are totally useless."

"Are you on to something?"

Josie jumped at the sound of Sophie's voice and quickly gathered her thoughts to respond.

"Yes, I believe we have found the problem. It is a software bug that only occurs when the components are put into the oven when they are already hot enough."

"Hmmm," Sophie intoned, "That seems to fit the pattern. But are you sure?"

Josie straightened and thought about how to verify her suspicion. Shortly, she described how they could verify that the **ovenPW** variable went negative and "rolled under" to 100% for components right around 145°. BJ found a blow torch and willingly heated a few components to her specifications, and as the calculated **ovenPW** approached 0% for the hotter components, it suddenly jumped from 0% to 100%.

Bug nailed!

To Josie's surprise and satisfaction, Sophie broke into a genuine smile and shook her hand and then Li Mei's hand. "I am extremely happy to find out that you two ladies knew what you were doing, and that you found and fixed this problem for us today. I think we will be able to make our deadline. No," she amended, "I am sure we will."

After they had verified the bug, it was quick work to change the declaration of the variable and test the new results again. Josie downloaded the new software into the oven's microprocessor and BJ fired up the line again. As the first hot component entered the oven, it was immediately ejected after the oven determined that it was already within the appropriate temperature range. She and Li Mei remained on site for the remainder of the shift to verify that components needing heating were also heated appropriately.

"So what did you think of your first onsite customer experience, Li Mei?" Oscar asked to start the status meeting. "You appear to have survived unscathed."

Li Mei's eyes widened perceptibly. "It was hard. I didn't know what to do at first. Sophie was not friendly."

"That's kind of typical," Josie interrupted. "Customers are already upset by the time you get there."

"Sophie is tough but fair." Oscar looked at both of them. "She was concerned that I sent engineers who have never been there, but you must have impressed her because she called me last night."

Oscar flipped his ID badge around the lanyard, clearly pleased. "She told me that you worked the entire time, and that she heard you brainstorming even before she was able to get you software listings or a computer."

A look of surprise spread across Li Mei's face.

"She told me that the next time something goes wrong, to send you two over again. That's high compliments from Sophie. You've just cemented a long-term relationship."

Ravi doodled on his pad and wondered when he would get his first site visit. Li Mei was barely here a month and had already been sent on the road.

Oscar met Ravi's eyes and then glanced away.

"Josie, do you have anything to add for the team?"

Josie set her pen down before she spoke. "Well, I guess we got sidetracked by the original problem report. It told us that the oven didn't shut off and that's what caused the failure. But that was never the problem. The oven never should have turned on in the first place."

"Hmm. That's a very important lesson for all of us." Oscar made a note in his logbook and glanced again at Ravi. "Why don't you summarize the brainstorming and bug tracking so Ravi can learn from the experience as well."

Another Page from Li Mei's Notebook

- *Make sure unsigned variables will not be used to store signed quantities.*
- *Check logic to make sure counting variables are properly bounded ("<=" versus "==").*

General Guidelines

- *Make an action plan. What do you know about product, customer, environment, hardware/software, materials used, priority, safety?*
- *List every symptom. Brainstorm what could cause each. The bug's location in the code can sometimes be determined before looking at the software.*
- *Classify the symptom.*
  - *One-time repeatable events - pattern to occurrence (e.g., start up, different behavior first time through function or feature).*
  - *Periodic - regularly repeating or occurs every time (e.g., tied to timer, interrupt, repeated calls to function/feature, hardware heartbeat).*
  - *Sporadic - seemingly random failures (e.g., boundary condition violations, unexpected input/output conditions, unhandled error conditions, faulty logic, hardware, timing, memory corruption and performance issues).*

Action Plan for Solving Problems

- *Interview problem reporter.*
- *Interview anyone who saw the system fail.*
- *Observe the system behavior - find out what is normal for the system.*
- *First identify the type of failure (one-time repeatable, constant, periodic, or sporadic) and then brainstorm possible root causes.*
- *Targeted search in software (or hardware) based on possible root causes.*
- *Hypothesize what should happen, and choose variables to watch.*

Chapter Summary: The Case of Thermal Runaway (Difficulty Level: Moderate)

In this mystery, an oven at Industrial Enclosures overheats and damages components. When Josie and Li Mei arrive, it is operational, but when it fails, it fails continuously. Josie introduces an Action Plan and symptoms for different types of problems using the temperature-controlled oven as an example. Although they aren't able to find the bug before it starts damaging more components, their thorough brainstorming allows them to nail it almost immediately after it finally reoccurs.

The Problem Symptom(s):
- Oven did not shut off, and overheated a component.

Targeted Search:
All HW/SW references to temperature are located in the software.
- Analog temperature sensor and reference read every second.
- Pulse-width modulation (PWM) routine controlling an external heater.
- Algorithm computing new PWM based on measured temperatures.
- Logic controlling timing in an interrupt service routine.

The Smoking Gun:
A component that was already at the proper temperature caused the oven to turn on 100%, when it shouldn't have turned on at all.

The Bug:
A simple casting problem caused this failure.
- Equation calculating the new oven PWM value produced a negative PWM value, which should have been truncated to zero to keep the oven off. However, the results were stored in an unsigned char, which converted the value of -2 into 254. The PWM algorithm truncated 254 to 100, turning the oven on at 100% and damaging the parts.

The Debugging Method Used:
- Interviewing people who saw the failure.
- Brainstorming root causes of reported symptoms.
- Isolating relevant functions and variables to watch during experiments.
- Comparing logged variables while the system runs, and again when it fails.

The Fix:
- Changed the variable **ovenPW** from unsigned char to signed char. (On memory-limited machines, using chars when possible saves valuable space. Depending on the bus and register widths, and on microprocessors architecture, operations on some data sizes are more efficient than others; for example, operations on chars can be faster on 8-bit machines than operations on ints. Choose your variable types accordingly.)

- Changed the logic to reset the incrementing **oven_pulse_width_ counter** from "==" to ">=". While this did not cause a detected failure, it was suspicious.

Verifying the Fix:

- Heated a component to the required temperature to verify that a) the oven did not turn on, and b) the calculation of **ovenPW** was properly truncated to 0%.
- Verified boundary conditions for **oven_pulse_width_counter** at 99%, 100%, and 101% to ensure erroneous values over 100 were properly handled.

Lessons Learned:

- Simple bugs can cause big expensive symptoms.
- Look for hidden boundary conditions. The PWM equation used an unsigned char - with a boundary condition at 0.
- Generate a debugging action plan, including the maximum time to spend chasing each idea.
- Don't overlook little clues - both failures occurred right after lunch with a new pallet of material.

Code Review:

The software is reasonably written. The variables have descriptive names, #defines are clear, and the code is self-documenting.

What Caused the Real-World Bug? In the baking Texas afternoon sun, the bias current to an unconnected processor input in the electronic billboard changed; the normal "zero" state drifted to a logic "one." Firmware didn't ignore this unused bit, and accepted the unnecessary and incorrect new data, generating a temperature display that was insane.

The moral is to do boundary checking. Anything over 150 degrees for this billboard is probably wrong. It would have been better to show nothing, or perhaps "HELP," rather than obviously silly results. [2]

References

[1] Simone, L. (2004), "A Feynman approach to debugging," *Embedded Systems Design* magazine, Vol. 17, No. 11: 20-31.

[2] Ganssle, J., Personal communications, August 1, 2006.

Chapter **5**

The Case of Two Inaccessible Microprocessors: Using Creative Methods to Understand System Behavior

Oscar returned the phone to its cradle and pondered a change in plans. He had scheduled a meeting this morning for the new Austin Home Medical Monitor project, and a team from B&L Omaha Telecom would be here shortly to review the requirements for the cell phone interface to the product. It was a big project they were designing from scratch and nearly the whole team was involved. But a frantic call about a machine failure out at Kelly Manufacturing added a headache to the mix.

He shook his head; he didn't understand why, all of a sudden, machines were flaking out in the field. This machine had been temperamental since the first time he worked on it, and he thought it had damn well picked a perfect time to throw another fit. He envisioned the six-foot mechanical arm hanging out onto the assembly line while boxes piled up behind it. Apparently, nothing could fix it but the big three-finger salute, although Oscar imagined that asking the customer to reboot the machine each time would not be well received.

Kelly Manufacturing had been a regular customer and he didn't want to screw up a good relationship. Someone had to get over there.

Today.

Exhausted from the late nights of assembling technical proof that Randy's latest pet-project idea wasn't feasible or cost-effective, he stewed and pulled up a spreadsheet of the team's open projects and scanned the deadlines. Ravi was assigned to a resurrected project they'd tabled last year but which now had a viable customer, and Li Mei was at the college taking a physiology seminar for her role in the Austin Medical Monitor project. And he had put Josie in charge of the technical interface with the Omaha team for the connectivity application between Austin and the cell phone.

Feeling a knot in his stomach, he spun his chair in a slow circle. The contents of his cube slid into view and then withdrew: a stack of customer requirements, trade journals, two half-empty soda bottles, reference manuals, his briefcase propped against the wall under several certificates of achievement. Then his Pink Floyd poster came into view and he stopped, staring at it; a simple prism hovering on a black background broke a single beam of white light into a rainbow of colors. The colors, sprung from nothing, reminded him of the new myriad responsibilities he'd never imagined. He was suddenly thrust into project scheduling meetings, negotiating for hardware support, approving technical content in marketing plans, and making sure he was available 24-7 for certain product launch sequences.

And the endless, time-sucking meetings with Randy.

What a load of crap.

He kicked off the base of the chair again, shaking his head to refocus on the Kelly problem.

That machine had run a few years without problems, but he remembered a service call several months back after label adhesive had built up around a carriage slide assembly, causing it to jam. There had been nothing wrong with the software; just crappy maintenance and a service manager with an attitude.

Who should he send? Josie was up for more responsibility, and that's why he'd assigned her to work with the Omaha Telecom folks today, to increase her visibility within Hudson Technologies. His gut told him Ravi wasn't ready for a solo service call. Ravi had potential, but he had trouble figuring out what to do next and needed more training to be self-sufficient.

Idly driving the mouse around the screen, he considered Li Mei. While she was new, she was learning to grasp the big picture. Josie thought she did well on her first on-site service call, and Sophie gave a positive report on both women.

But, he conceded that this service call required some experience.

Making up his mind, he found Josie in her cube reading the latest issue of *Embedded Systems Design* magazine.

"Remember that box-labeling machine over at Kelly Manufacturing? They are complaining that it runs for several hours and then stops running. They reboot it and it works for another couple of hours. Can you head over there now and take a look at it?"

"You said that machine was a mess." She folded back the corner of a page and asked, "And what about Omaha Telecom?" Oscar watched as she carefully balanced the magazine on the pile of unread trade journals. She had the same precarious pile as he did; each one read was quickly replaced with three new ones.

"All that machine does is move a box into position and then stamp a label on it or something like that, right? How can it keep failing?"

"I don't know. It's not that complicated." He hoped it was just another dirt issue caused by the idiot in their maintenance department. "Maybe you can get back here in time for the Omaha Telecom meeting."

Ravi shoved aside another box and opened the one below it, looking for the missing emulator. Pulling out power cables, a keyboard, and several tangled mice, he found himself at the bottom of the junk pile with no luck. Seven other boxes lay open behind him.

The emulator was nowhere to be found.

This is just perfect, he thought. He'd looked everywhere, and no one could remember where it had been stored. After another twenty minutes of poking around under lab benches, he gave up.

At his bench, Ravi took inventory of the pile of hardware he'd collected. From the mess he extracted three prototype boards. Each had a microprocessor and a FLASH chip for program and memory storage, a display, some pin connectors for ribbon cables that ran to a separate motor board assembly, and a port for debugging.

The port where the emulator would plug in, if he could find it.

Picking through the pile of cables, he extracted what he thought was a complete set of hardware and shoved the rest out of the way. He read the bug report again.

"CR Defect #2005011372 Description: The digital clock display on the RoboGym Control Panel Module doesn't run smoothly. The numbers don't update at the right time. Severity: 1 (highest) Priority: 1 (highest). Additional Information: It's the one with the blue label, not the red one."

His frustration simmered into anger as he realized the Change Request was classified as Severity 1, which meant that someone would be pressuring him very soon to find out why it wasn't fixed already.

He had no clue how to start fixing it without an emulator.

Josie's folder on Kelly Manufacturing was thin. The company had a diverse product line and used a fair amount of automation to combat expensive local labor rates. Squinting into the morning sun, she nearly drove past the squat building on a tiny lot right off Route 287. She maneuvered around delivery trucks in the parking lot and pulled into an empty spot between a black Corvette and silver BMW. She peeked in the window of the Corvette and admired the interior, idly wondering what kind of salaries Kelly was paying.

She found the factory floor in full swing, automated machinery humming and conveyers carrying boxes and parts from one area of the large open expanse of the factory floor to the other. For all the activity, it was surprisingly quiet as a thin, unhappy man explained how the machine had been behaving.

"It's been doing this for a couple of weeks now. It runs great and for no reason faults out. I don't know what's wrong with it, but a red light goes on in the control panel."

Josie listened to Larry's description as they watched the machine pull boxes into its maw and then spit them back out again. The machine was a ponderous thing that automatically extracted single boxes from an assembly line, measured and weighed each, and then applied a variety of product and shipping labels.

The machine was conforming to the known Laws of Embedded Systems: it knew when it was being watched, and it behaved accordingly. Little boxes flowed in and out; it was running flawlessly.

"Has it failed yet this morning? I'd like to see which red light goes on." She hiked her bag over a shoulder and ticked off in her head what a red light could mean. Maybe a motor controller or position sensor out of range.

"Hasn't failed yet today." Larry grabbed her arm and yanked her forward as a forklift screamed past them. Josie jumped off balance in surprise and glared at him. The forklift really hadn't gotten that close.

"Gotta talk to that guy," Larry muttered, looking back over his shoulder as the forklift disappeared around a corner. "Sonny is a menace on two wheels with that thing."

Josie smoothed the expression on her face and looked around. "Is there some-place I can hang out and wait for it to fail? I'd like to be close by." Larry led her around the side of the machine and pointed to an area crosshatched in yellow, then left before she could ask to see the control panel.

Helpful guy, she thought sarcastically. Well, hopefully the machine would fail soon.

She looked back at the machine and settled in to wait.

The sound of the machine and the light bouncing off the moving parts was rhythmic, almost hypnotic. All of the boxes coming down the assembly line were about the same size. Occasionally a smaller box came through. The carriage arm swept each off the main assembly line slightly up into the machine, and then shortly after, the arm would travel out in the reverse direction, allowing the box to roll back onto the main line.

Back and forth. In and out. Each box appeared like it might overshoot the assembly line and roll off the other side, but none did.

She wondered what could be wrong, and thought about the brainstorming conversations she had had with Li Mei. This was like waiting for that oven to fail, she realized.

A red light could mean that an error condition was detected. Maybe she could access an error message or find out what part of the system was failing, but at the moment, she had little information to guide her thoughts.

The morning dragged on as she finished a cup of coffee she'd snagged from a break room. As she drained the last of it, the carriage stopped and a buzzer sounded.

Looking around for Larry, she saw him tromp back to the machine while yelling into a walkie-talkie. His scowl was acknowledgement enough that the machine had failed.

Finally! She tossed the empty cup in a nearby trash bin and met him at the carriage.

Larry shoved the walkie-talkie back into his belt clip and gestured at the hovering carriage arm. "This is what the damn thing does." Josie waited for him to add something useful, but he just looked around until a heavyset man close to her father's age ambled into view. "Don will show you."

The man called Don cocked his head at her and chortled.

"Well, look who they sent."

He paused to look her over completely, and Josie felt herself go on yellow alert. Great, she thought, another sleazebag, and suddenly wished that Oscar had taken this service call instead of her. Surreptitiously, she located the nearest exit and checked her surroundings.

Don was already walking to the machine, gesturing to the hovering arm.

"The carriage is pushed all the way out and ready to draw in the next box. It goes out far enough, but it's supposed to rotate this little foot bar around behind the box. But it don't. It just hangs there stuck."

He wandered around to the back of the machine and yanked open the control panel. It screeched and clattered open. Josie sighed and followed without comment, pulling her long hair into a messy ponytail. Looking around the mostly unfamiliar panel, she identified the motion controllers, motor drivers, optoisolators, power lines, and computer control. A bank of solenoids along the side controlled a river of pneumatic tubing entering and leaving the panel.

Everything was coated with a layer of grey dirt, which she noted was probably a direct consequence of the missing fan filters on the side of the panel. She wondered what happened to the filters, and was surprised that they could even see the LED indicators. Well, she amended, the red indicator Don pointed to was glowing fairly strongly under all the grime. The indicator belonged to a set of screw terminals that appeared to be input or output lines from the computer controller.

She nodded to indicate she saw it, and then asked, "Why'd you take those filters out?"

"They were clogged."

Biting her tongue, she choked back the obvious response.

"Anyway," he continued, "This is the light. Same one each time. So what's the problem?" Don crossed his arms and fixed her with a leering smile. The sallow skin on his face sagged, as if he had once been even heavier but had lost a lot of weight. He brushed greasy hanks of hair out of his eyes. The guy creeped her out.

"I'm not sure yet." Josie shrugged and feigned nonchalance. She continued to look around the control panel, wiping off selected components with a napkin, and then pointed to the glowing light. "This looks like it's an indicator for an input port. Can I see the software and documentation?"

"Larry's getting it." Don had pulled up a stool and sat, arms crossed.

Shortly, Larry returned with a laptop, which he handed off, and immediately left again. She set it on the top of a nearby table and prepared to immerse herself in the software.

When she opened the file, she groaned (Figure 5-1). It was written in a motion-control language she didn't know. She hoped the code was well documented.

> **Reader Instructions:** You may be unfamiliar with this motion-control language, but it is a straightforward language to learn. The main loop is bounded by the "def main" and "end" statements. Sketch a flowchart for the main loop.

```
Digital Inputs to Port 1
; 1IN.17        User Input: Machine Start Cycling Button
; 1IN.18        User input: Machine Stop Cycling Button
; 1IN.19        Box proximity detector
; 1IN.20        Linear Carriage Max Limit Sensor
; 1IN.21        Rotation Carraige Max Limit Sensor
; 1IN.22        Linear Carriage Min Limit Sensor
; 1IN.23        Rotation Carraige Min Limit Sensor
; Max X sensor (input to limit port - no code requried)
; Max Y sensor (input to limit port - no code required)

SCALE1                  ; Allow scaling stepper motor counts to inches
SCLD 25000,1            ; Distance scaling values for motor 1, motor 2
1MA0                    ; Axis 1, move incremental distances (not absolute)
2MA0                    ; Axis 2, move incremental distances (not absolute)

del main
def main
   gosub inits          ; perform all initializations

   ; Wait for user to press START
   l0
     if (1IN.17 = b0)
        lx               ; Exit loop
     nif
   ln

   ; Loop here forever processing boxes
   l
     if (1IN.18 = b0)    ; Check if user selected Machine Stop
        lx               ; Exit loop
     nif

     ; Move carriage arm out
     1 D 56             ; Set distance to move out 56 inches
     1 GO 1             ; Execute the motion
     wait (1as.1=b0)
     T .5                ; Pause
     ; Rotate carriage arm down behind box
     2 D 90             ; Set distance to 90 degrees
     2 GO 1             ; Execute the motion
     wait (2as.1=b0)
     T .5                ; Pause
     ; Wait to detect box at carriage arm
     wait (1IN.19 = b0)
     ; Move carriage arm back in with box for labeling
     1 D -56            ; Set distance to pull arm back in 56 inches
     1 GO 1             ; Execute the motion
     wait (1as.1=b0)
     T 1.0
     ; Weigh and measure box, and then apply the labels
     gosub weight
     gosub labels
     ; Move carriage arm back out to the assembly line
     1 D 56             ; Set distance to move out 56 inches
     1 GO 1             ; Execute the motion
     wait (1as.1=b0)
     T .5                ; Pause
     ; Rotate carriage arm up from behind box to release it
     2 D -90            ; Set distance to 90 degrees
     2 GO 1             ; Execute the motion
     wait (2as.1=b0)
     T .5                ; Pause
     ; Retract carriage arm back into the machine
     1 D -56            ; Set distance to move back 56 inches
     1 GO 1             ; Execute the motion
     wait (1as.1=b0)
     T 10.0              ; wait for box to leave area
   ln
end
```

Figure 5-1 Software Listing for the Kelly Labeling Machine.

Scanning the top of the file, Josie found the hardware declarations and references to two motors, several banks of digital inputs, a bank of digital outputs, and a bank of analog inputs. She quickly identified that Port 1 received inputs from a number of digital sensors. The machine had two carriages, each controlled by a separate motor. A quick trip back to the control panel told her that the glowing red light was connected to a limit sensor on the linear carriage arm, on input line 20 of port 1: "1IN.20".

She looked up and called to Don. "There's a limit sensor on the carriage, something that detects if it moves out of range?"

"Yup, there's a bunch of sensors out there, all right."

Josie walked to the carriage arm and noted that it was mounted on a belt-driven linear actuator that was extended about five feet, nearly to the end. (See Figure 5-2). Looking around where it attached to the main frame of the machine, she noticed a proximity sensor that was mounted to the frame. She squatted and looked up to see a small metal bracket bolted to the bottom of the carriage; it was located nearly over the top of the proximity sensor. Sighting down the beam, she saw that a motor was mounted on the other end.

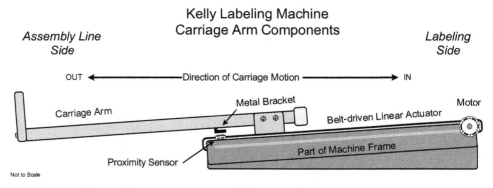

Figure 5-2 Kelly Machine Carriage Arm Components.

Josie chewed on her lower lip, and then realized that the proximity sensor must be magnetic. It must have signaled the motor to stop when the metal bracket passed over it. She estimated the air gap between them at maybe a quarter of an inch.

Looking over the relative position of the parts, she decided that the motor controlled the linear motion of the carriage and that the carriage had somehow traveled too far along the actuator, and then had gone out of range because it had stopped with the proximity sensor right over the metal bracket. The sensor tripped, causing the motion controller to stop the motor and halt the machine. On closer inspection, she noticed that the proximity sensor had a small indicator light inside. That indicator was also lit.

That clenched it, she thought; the carriage arm had gone too far. She hadn't even needed to read much of the code to get this far, but seeing the light in the proximity sensor boosted her confidence.

Standing up, she looked around and found Don still sitting in the same position, watching her. She hollered to him as she walked back, "It looks like the carriage traveled too far out of the machine. It tripped the prox sensor here and the motion controller halted the motor."

She hesitated, unsure if she wanted to engage him in prolonged conversation, but she wanted to be sure that the problem was repeatable before she started debugging. "Is this what happened last time? Did the carriage stop in this position?" If this was really an out-of-range problem, it could be hard to find.

To her surprise, Don actually paused to think, and then gave her what sounded like a straight answer.

"Well, I didn't check the limit sensor every time, but I do tend to recall that the carriage was extended before. We had to restart the machine to get the arm off the main conveyer belt because boxes started piling up behind it." He glanced at his watch. "Can we turn the line back on and let it run while you think about it? We got production to make."

"I guess so. But can I connect up a debugger to see what happens right before it fails next time? It's hard to know what's causing this without looking at the variables."

Don shook his head. "No can do. You can't be back here when the line is operating. The control panels have to be closed."

Josie groaned and counseled herself against rolling her eyes. He wanted her to fix this using ESP rather than concrete debugging tools. Typical manufacturing plant - "Fix the machine, but we can't afford any down time." She shrugged her assent, grabbed the laptop and walked back to the yellow crosshatch area.

Shortly after the machine resumed normal operation, Don wandered off.

Good riddance, she thought. If it weren't for this machine, she would be at the Omaha Telecom meeting. Missing the meeting upset her. The more senior developers got invited to customer meetings and were then rewarded with their own parts of new development projects. And she was stuck here with a temperamental behemoth and its two unpleasant keepers.

Around her the sounds of the factory droned and buzzed, with an occasional reverberation of banging steel. Fresh air from two open bay doors filtered through the plant floor. Segregated against the wall, she perched on the edge of an empty wooden crate and scanned through the software listings, listening to forklifts squeal in the distance. Somehow she would have to figure out how to debug a problem without actually being connected to the machine.

ESP, indeed.

Oscar had had nearly enough.

Josie wasn't allowed to connect the debugger while the machine was running, and she vented that the maintenance guy was the same idiot he had warned her about. And then Ravi had shown up angry that he couldn't find the magic emulator for the Robo Exerciser whatever-it-was and that the CR was top priority, and demanded to know how he was supposed to do his job.

Between their complaints, and the people from B&L Omaha Telecom being missing in action, nothing was getting accomplished. It was already 11:11 in the morning.

He threw up his hands. For this he became a technical manager? With annoyance laced with an inchoate sense of duty, he trudged off to find Ravi. He was flabbergasted that neither of his team members showed any creativity in dreaming up a different way to debug their problems.

As Ravi had described it, the project was practically unmanageable.

Without preamble, Oscar told him to fire up one of the RoboGym prototype boards and see what it did. To Ravi's surprise, it powered up with an all 8's test on the LCD display and then began blinking 12:00:00.

"Well," Oscar said dryly, "it looks like part of the display is a clock or a timer. Perhaps you should try setting it."

Ravi seemed surprised that he could easily program values for hours, minutes, and seconds.

"The CR says that the numbers don't update at the correct time. It doesn't run 'smoothly'." Ravi hovered over the prototype. He was edgy, and Oscar noticed as he fidgeted that it looked like Ravi hadn't slept much lately.

Oscar stood beside him and waited. Counseling himself not to flip his ID badge around its lanyard, he invoked one of his favorite debugging methods; just sitting and watching the system.

Patience.

Josie was relieved that the code contained comments, although she knew that sometimes the comments themselves became buggy if they weren't updated along with software. She readjusted her leg, which had started to fall asleep, and scanned the code for the software that serviced the proximity sensor.

She verified that the proximity sensor that had tripped was indeed 1IN.20. Digital input line 20 on Port 1. Since this appeared to be the linear carriage, she guessed that the rotation carriage was the little foot that Don described as dropping behind the package to pull it in for processing.

But nowhere else in the code could she find where the linear carriage port 1IN.20 was referenced.

Capturing a section of hair that had escaped her ponytail, she hooked it behind one ear and bent back over the computer.

She commanded herself to begin again, at `main`.

One line at a time, she interpreted the function and its meaning and moved on, not chasing a subroutine or function, not referencing a hardware port or macro. If the main routine was properly written, it would speak to her about the overall function of this product. She began reciting to herself, unaware that she had begun to tap her foot to the background machine rhythm, the heartbeat of the factory.

"First, everything gets initialized. Then the system waits for the user to press start." She wasn't sure of the exact syntax for a loop but decided to trust the comment for now.

"Next," she intoned, "there is a big loop." She paused to look at the main actions in the loop, and then ticked each off on her fingers.

"One, check if the user turns the machine off. Two, move the carriage arm out 56 inches so it's behind the box. Three, rotate the arm down 90 degrees. Four, wait until the box is detected at the carriage arm using a sensor on Port 1 pin 19." She made a notation on her pad to check the input port pins and sensors for this logic. "And when the box is detected, bring the carriage all the way back 56 inches, and then weigh and label the box. Finally, put the box back on the conveyer line and get ready for the next box."

Josie exhaled. That was it. The basis for the rhythmic cycle of this machine (Figure 5-3).

Now, she wanted to see where it could fail. She scanned backwards through the small bits of software controlling the main functions and stopped at the code fragment that initially moved the carriage arm out. Picturing how the carriage was

Kelly Labeling Machine Processing Loop

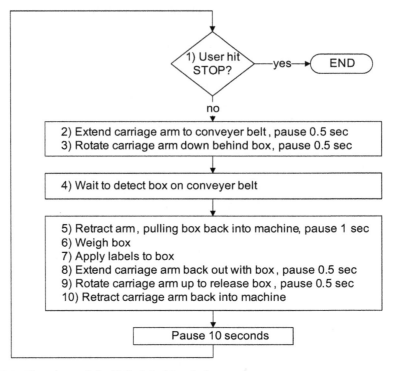

Figure 5-3 Flowchart of the Kelly Machine Software.

positioned when it stopped, she imagined program control hanging on the wait statement, but the format of the line of code baffled her.

```
wait (1as.1=b0)
```

Looking at the computer system tray, she realized she had wireless connectivity and she cut-and-pasted the line of code into a search engine. As she waited for the Programmer's Guide PDF to download, she guessed that the numbers preceding each command statement referred to the two motor axes: '1' for the linear motor carriage, and '2' for the rotation arm.

A squeal of tires made her look up, and she saw the forklift round a corner and charge toward her. The same guy was at the controls, a young man with a deep tan and happy grin. His smile was infectious. When he was sure he had her attention, he raised the forks and zipped around the corner on two wheels, winking as he passed. Josie laughed as she watched him speed away.

Thankfully, the factory floor had a strong wireless signal and the download completed. Now she could reverse-engineer individual lines of code more easily.

The RoboGym module was most certainly NOT running according to spec, Ravi concluded with dismay.

As Oscar peered over his shoulder, he nervously watched seconds tick by as his simulated "workout" progressed; a steady stream of digits from zero to fifty-nine. But, he noticed with a spark of curiosity, each time the counter reached 59 seconds, it paused, seeming to hover in a short void, prolonging the instant of time, just long enough to be noticeably out of spec, before toppling over to 00 seconds and then continuing in the orderly periodic march of one second after another as time moved on.

The pattern was being broken on a regular basis.

An idea solidified and he grabbed a stopwatch, wondering if the clock kept accurate time. After hovering over the clock and stopwatch for several minutes, he found that every elapsed minute was a true minute, which meant that the longer 59th second was compensated for by a shorter 00 second. This was useful information, and he stood to propose a debugging plan to Oscar.

Oscar preempted him. "Ravi, last time you had an emulator to measure function times. This time, you don't have one, but you also don't *need* one. You have the tools you need right here in this lab."

Ravi felt his jaw drop, and quickly closed his mouth. "But how do I step through the code to find out what's wrong if I don't have an emulator or debugger?"

"Weren't you paying attention at the team meeting when Josie explained how she taught Li Mei about periodic and sporadic events?" Oscar fixed him with a stare. "And just what does 'smoothly' mean, anyway?"

"I just measured that the clock is okay, just those two seconds are wrong!"

"Define 'wrong.' And how accurate is your thumb on that stopwatch? How did you know when to start the measurement, and then stop it? Does this happen every hour, too?"

Without a pause, Oscar changed gears. "You made a start. You're an engineer; now you figure out how to prove what works and doesn't work with that module." He asked Ravi, "Do you have the compiler?" Ravi nodded and Oscar concluded, "If you can load new code, you don't need a debugger or emulator."

Ravi stood speechless as Oscar turned and left the lab.

> **Reader Instructions:** Ravi's idea to characterize the timing of the bug is good. Why didn't Oscar like his method? What other ways can Ravi access the code's behavior in real time without a debugger or emulator? Think about the specific problem. What types of functions should he look for in the software when he sees it?

Ravi returned from lunch to check the elapsed time clock and stopwatch experiment that he had set up nearly an hour ago. He'd programmed an hour-long workout, and shortly the elapsed time on the stopwatch reached one hour, rolling over from 00:59:59 to 1:00:00. Right beside it on the bench, he saw the clock pause slightly at 00:59:59 before rolling over as well. Nodding his head in satisfaction, he concluded that this test was a much better confirmation that the clock was still accurate. If each elapsed minute added a half-second delay, he should have observed that the time displayed on the two devices slowly drifted apart, with the clock reaching the one-hour mark about 30 seconds slower than the stopwatch.

He debated going to tell Oscar, but opted instead for a short email update before scanning the software listing for anything that looked like a clock function. Before he was past the first page, thoughts of his previous lesson with Oscar came flooding back and he grabbed a notepad instead.

"'The display doesn't run smoothly' is what the CR said." Ravi wrote "smoothly" in block letters on the pad and thought about what he had observed. The person who wrote the CR wasn't as thorough as Eduardo generally was, but the information was correct. This was a periodic event as Josie had described, and it happens every one minute, just as the clock rolled over.

What caused periodic events? He racked his brain to recall the team meeting discussion that had ensued. Li Mei had written a list of causes on the whiteboard. Bugs in a function called periodically. Or in an interrupt. An incrementing variable rollover. Or something external to software like jitter, or a charging and discharging capacitor. Even debugging equipment can interfere with timing and program execution. Although, he realized with annoyance, the problem wasn't caused by an emulator since he couldn't *find* one in the first place.

After some thought, he decided to go with an obvious source of error - whatever periodic subsystem controlled the clock. He jotted some of the normal places in a software program where period timing originated: "clock subsystem," "periodic timers," and "ISR." Reasonably more confident, he turned back to the software listing (Figure 5-4) and quickly found a logical starting point - an interrupt service routine: **void TIMER_ISR(void)**.

> **Reader Instructions:** Based on what Ravi has learned so far, what parts of the software could cause the observed problems? How would you prove it? (Note: if source code is not provided for a function, assume that the code exists and that the function works correctly.)

```
unsigned char counter_200_msec;
unsigned char seconds, minutes, hours;
unsigned char LEDControl_f;

/* ----------------------------------------------------------------------------
Name:       TIMER_ISR()
Function: Interrupt service routine.
---------------------------------------------------------------------------- */

void TIMER_ISR(void)
{
    /* ------- 200 msec ---------- */
    counter_200_msec++;
    if (LEDControl_f == LED_ENABLED)                  /* Logical A    */
        update_LED_state();
    update_sensor_control();

     /* ------- 1000 msec ---------- */

    if (counter_200_msec == 5)                        /* Logical B    */
    {
        counter_200_msec = 0;
        seconds++;

        if (seconds == 60)                            /* Logical C    */
        {
            seconds = 0;
            minutes++;
            sensor_params();

            if (minutes == 60)                        /* Logical D    */
            {
                minutes = 0;
                hours++;
            }
        }
        updateClockDisplay();
        update_motor_profile();
    }
}

/* ----------------------------------------------------------------------------
Name:       updateClockDisplay()
Function: Create ASCII string for clock time (hh:mm:ss) and send to display.
---------------------------------------------------------------------------- */
void updateClockDisplay(void)
{
    convert_clock_to_ascii(time_string, hours, minutes, seconds);
    write_clock_time(time_string);
}

/* ----------------------------------------------------------------------------
Name:       sensor_params()
Function: Based on sensor readings, dynamically recalculate coefficients.
---------------------------------------------------------------------------- */
void sensor_params(void)
{
    acquire_location_information(&radius, &x_location[0]);

    sensor_coefficient[1] = multiply (radius, (long) x_location[1]);
    sensor_coefficient[2] = multiply (radius, (long) x_location[2]);
    sensor_coefficient[3] = multiply (radius, (long) x_location[3]);

    sensor_coefficient[1] = divide ((long) x_location[1], 5.5);
    sensor_coefficient[2] = divide ((long) x_location[2], 5.5);
    sensor_coefficient[3] = divide ((long) x_location[3], 5.5);

    update_hardware_values (sensor_coefficient[1], sensor_sensitivity[1]);
    update_hardware_values (sensor_coefficient[2], sensor_sensitivity[2]);
    update_hardware_values (sensor_coefficient[3], sensor_sensitivity[3]);
}
```

Figure 5-4 Software Listing for the RoboGym Clock Timer.

Ravi studied the code and the sparse comments, and began to sketch a rough flowchart of the interrupt service routine. Every 200 milliseconds the ISR performed some LED and sensor functions and incremented a counter. When that counter reached 5, it incremented another counter called **seconds**. He mulled over the cascade of logical statements and incrementing counters before finally convincing himself that the **seconds** variable actually did increment every second.

It was a good sign to find a variable named **seconds**. It might point him to the source of the problem. Could that variable actually be the one used to update the display? Scrolling down in the file, he found a function called **updateClockDisplay()** that did, in fact, use the hours, minutes, and seconds variables that he had found in the ISR. He felt his excitement rise as he traced the **seconds** variable till it reached 60 and was reset to zero, and then the minutes variable was incremented. In the same manner, **hours** was also incremented.

He groaned. Nothing looked out of place.

Returning to the ISR code, Ravi tried to identify the exact line in the function that corresponded to what he observed on the clock's display. He placed the index finger of one hand on the line that incremented the **seconds** variable, and the other on the **updateClockDisplay()** function.

It's one of these two, he decided. Where **seconds** gets incremented, or where it gets displayed.

Or, maybe the problem lay with how the variables were displayed. He searched through the function used to convert the variables to a time string and then, with some reluctance, began digging through hardware driver code in search of the bug.

As he listened, Josie explained again how the carriage arm was extended when the Kelly machine failed.

"Reading the code," she continued, "I think it gets stuck at the wait line after the command to move the carriage out 56 inches. Am I reading the code right?"

Cradling the phone between his ear and shoulder, Oscar scanned a printout of the code and started to explain the notation. Josie interrupted him. "I found that online. '**1as**' is the status on motor axis 1, and status bit 1 after the decimal tells me if the motor is still moving. That part makes sense; it waits for the carriage to extend all the way out and stop before rotating the foot behind the box. But how can it get stuck right there? The motor is *clearly* not running. The program obviously hasn't made it to the next couple of lines that command the second motor because it stays off." After catching her breath, she continued, "It has to be stuck *right there*, and this jerk-of-a-maintenance-guy won't let me hook up the debugger."

Although Oscar heard her frustration, he smiled and imagined her looking back over her shoulder as she said it. Jerk or not, wandering around the manufacturing floor while automation was running wasn't exactly safe.

"Wait a second, Josie. What about the indicator light? Didn't you tell me that the machine is throwing an error? Couldn't that be what's stopping the motor?"

"Oh yeah," she conceded. "I forgot about that."

"So if you put the debugger on that line of code, it may not buy you anything. The machine will go ahead and fault out and the debugger will stop on assembly code in some function you can't access." He stopped to think about how she could debug the problem. These motion-controller packages had a lot of prewritten software routines to handle error conditions.

"It's the Carriage Max Limit Sensor, on input port 1 line 20," she offered.

Oscar scrolled up to the port definitions. Something wasn't right with the software, but he couldn't get a handle on it.

"Josie, let me call you back. And email me that software code file."

After sitting nearly a minute, motionless, he felt the haze starting to lift. Without realizing it, his eyes slowly refocused on the paragraph pinned to his cube behind his monitor and he let the words flow over him.

> "The most important debugging tool is a healthy dose of common sense, a sometimes rare commodity in the frenzy of debugging . . . . Part of the challenge of debugging embedded systems is in out-thinking the program and devising new means of getting useful information out of the limited I/O found in the system. Be creative." [1]

The Kelly labeling machine and the RoboGym module problems weren't so different after all.

Oscar's hands flew over the keyboard. He and Josie needed to Be Creative.

Just as he had instructed Ravi.

Hitting redial, he waited for Josie to answer and then began without preamble.

"You don't always have all the perfect tools available, or in your case, you have 'em but you can't hook 'em up." He attached the updated software code listing to the email message and hit SEND. "So, we're making a new debugging tool. I just sent you new code. Load it up."

"What does it do?" Josie asked.

"This could be pretty cool, but I don't know if it will work. Did the speaker on the machine buzz when it failed?"

"Yes."

"Good. The speaker needs to be functional for this." Oscar hoped it would work. "The software tells the motor how far to travel but we don't know if it actually does what is commanded. We can't measure it, and we can't look at the software.

"BUT!" he added with flair, "we can ask the software nicely to tell us."

He waited until Josie found Don and got approval to stop the line and load the new software. When she returned to the phone, he asked her to listen carefully after the machine started processing boxes again.

After a pause, Josie relayed, "It just finished a box and the machine made two separate beeps. Is it supposed to do that?"

"Yes. Keep listening while it does the next box."

"Okay, two more beeps."

"Do the two beeps sound the same, or are they different pitches?"

"I'm not sure; are they supposed to be different?"

Oscar debated telling her and opted for partial disclosure. "I sent two tone commands to the speaker. The first tone represents the distance the motor is commanded to go. 56 inches. The second is the actual motor encoder position after the carriage arm has extended all the way out."

Silence. He pulled up the software listing to confirm his suspicions while he waited for the machine to sing.

"Oscar? Still there?" Josie's voice crackled.

"Yup. Go ahead."

"I think the second beep changes each time the arm extends out. The first stays the same." She paused. "This machine has stepper motors; I thought encoders were to control position for servo motors."

"You can use them on stepper motors, too, although when steppers run correctly, just counting the pulses you command is often good enough. If you need more precise control of position, you'd wait until motion stopped, read the encoder counts to see how far the motor *really* went, subtract the two to get the error counts, and give the motors another command to make up the difference. For this machine, we are commanding the motor to go 56 inches, but a couple of encoder counts of error is trivial."

"So you made 56 inches like 56 Hz or something?"

"Not quite - that frequency is pretty low for human ears to hear. I just hacked together a little equation to translate numbers representing position into pitches or tones, but I chose a starting pitch a little higher in the audible frequency range. The value for 56 inches is about 261 Hz - that's middle C on the piano." Oscar rushed ahead, "So what does a change in pitch or tone tell you?"

"Well, if tone equals motor position, and the second tone seems to go up relative to the first a little bit for each box," she paused, and then finished in a rush, "then this means for each box processed, the carriage extends a little further out each time!"

"Bingo. And after a while, the carriage reaches the max position and trips out the max limit sensor. So what could cause the motor to do that? I'll give you a hint: the tone is the actual position, not the commanded position."

She hesitated again before answering. "I don't get it. If the motor missed pulses, it would go a shorter distance. How could it go a longer distance?"

"But it *is* going a shorter distance than commanded." He paused to let his words sink in. "Not when the carriage arm goes *out*, but when it brings the arm back *in*."

"Oh, I get it!" Suddenly Josie's voice changed from patient frustration to elation. "The boxes are too heavy! When it pulls the box up into the machine, it loses pulses and doesn't bring it all the way into the home position. Then, when it lets the box roll back out to the assembly line, it goes its commanded 56 inches, which takes it a little further out than it's supposed to."

"That's what I suspect. Verify it, change the software so that it returns to the home position each time instead of just reversing the 56 inches, and then get back here, okay?"

* * *

"Here's what you are going to do. A slightly different debugging technique." Oscar grabbed the RoboGym module from Ravi's hands, flipped it over, and began unscrewing the case. After he had exposed the circuit board, he leaned across the bench, dragged the scope probes closer and then peered at the board.

"This is the microprocessor right here." He pointed out the integrated circuit to Ravi. "Unfortunately, the IC is surface mount, so it's hard to clip the scope probes right to the leads."

"What are we going to measure?" Ravi asked him.

"We are going to wiggle some pins." He motioned for Ravi to follow, and flipped on a soldering iron nestled on a bench littered with tools and bits of wire. "I think you might be right about the interrupt service routine. There's a hiccup in there I want to isolate." He added with some annoyance, "The clock update method appears to be less than robust."

After the iron came up to temperature, Ravi watched Oscar, who glanced at a schematic he had brought and then bent over the board to work.

"I found a spare pin on one of the output ports. Port A pin 5. Nothing is using it." Smoke curled over his head as he tack-soldered a green wire-wrap wire to one of the microprocessor pins. "I am attaching a wire to that port pin, and another to ground." With a flourish, he shot the soldering iron back into the holder and flipped it off. Ravi looked at the green and black wires and wondered what obscure debugging method Oscar had decided to use.

Back at Ravi's bench, Oscar connected a scope probe and ground clip to the wires. "Without a debugger or emulator, you can't see into the processor like you are used to. But for this problem, you don't need that kind of total visibility. You just need to check timing of different parts of the code." He pointed to the port pin he had soldered to. "We will change the software slightly to configure this pin as a digital output, and then write 1's and 0's to the pin. Then we can see what's going on in the software from the scope display."

"But won't adding software change the behavior of the system?"

"Ah, yes! You are right - behavior can change, but we are going to add very little bits of software that will give us very valuable information about timing inside the processor." Oscar grabbed a notepad and printed a snippet of code.

```
PORTA(5, HI);
barney();
PORTA(5, LO);
```

"This command sets pin 5 on Port A HI or LO. **barney()** is just a random function." Then he asked, "If we hook the scope to this pin, what would we see when this runs?"

Ravi pondered. Shortly, the answer came to him.

"We would see a rectangular wave on the scope. It would go high right before **barney()** executed, and return low right after. That would tell us when the function **barney()** was executing." He looked up expectantly, but Oscar's expression told him he hadn't quite nailed it.

Oscar prompted, "Why are we using a scope? If all I wanted to know is when it ran, why didn't I just hook it up to a buzzer or something?"

"Oh, we are trying to measure time." Ravi looked back to the code, and then realized what Oscar was trying to show him. "We use the scope to measure how wide the pulse is. That tells us how long the function takes to execute." He thought about how he could use this pin-wiggling method, and turned back to his computer.

With Oscar forgotten for the moment, he edited the ISR to add the pin-wiggling commands before and after **updateClockDisplay()**.

```
PORTA(5, HI);
updateClockDisplay();
PORTA(5, LO);
```

He quickly had the code recompiled and downloaded into the microprocessor. After resetting the system, he looked expectantly at the scope and was rewarded with a periodic rectangular wave that marched across the screen.

"Wow - look at that. It worked!" He turned to Oscar, grinning. "This measures the time it takes to update the time on the display."

"Good job. Now what?"

"Well," Ravi leaned in to the scope and hit PAUSE to capture a screen's worth of data, and then fiddled with the controls to invoke the screen cursers and manually measured the width of the pulse. After a moment, his hopes fell. "The pulse width is only about 30 milliseconds."

"You only measured one pulse. You already know that 59 out of 60 time updates are okay; maybe you just captured one of the good ones."

Ravi let the system run again, careful to make sure he observed one of the delayed rollovers before he stopped the scope. Scrolling back through the data-capture memory, he found all of the pulses were about the same width. None were close to the several hundred milliseconds of extra time they were searching for.

Oscar broke into his train of thought and asked, "Why are you not investigating *this* red flag?" He leaned in to the monitor and pointed. "What the heck does `sensor_params()` have to do with a one-second timer?"

"Nothing, I would guess. So why look at it?" Ravi was confused.

"What does that function do?"

Ravi scrolled down. "Ah . . . it does some math to update some hardware parameters. I don't really know what all that is."

"You don't have to understand its function, but you should recognize that it has a slew of time-intensive functions and casting operations." Oscar looked at him and shrugged. "You think these are good things to have in an ISR?"

"I guess not, but maybe the author of this code thought these needed to be updated every second." He amended his train of thought and turned back to the keyboard, "Let me check how long it takes to execute that function."

Soon he had a new pulse train on the screen and after quick work with the cursors, announced, "They are not all the same, but they are all about 300 - no, wait, 290 milliseconds each."

Oscar whistled low. "That," he added after a pause, "appears to be the proverbial smoking gun. The ISR is supposed to run every 200 milliseconds."

Mulling over the results, Ravi added, "Yes, that could be it."

They both sat in silence. With Oscar flipping his ID badge and staring into space, Ravi decided to run a quick test to see if temporarily commenting-out the offending function would make the clock run smoothly again. It took only moments to check, and shortly he was hovering over the clock display waiting for the pregnant pause.

The clock counted to 59 seconds and, with the sharp precision of a marching band, stepped smartly to 00 seconds without a hitch.

That was it.

"Looks like you found the problem." Oscar stood and brushed invisible crumbs from the front of his shirt. "I've got to check in on Josie - she's got a nonstandard debugging challenge as well. In the meantime, you've got a big architectural flaw in this software. The ISR is periodically running over 100 percent, which means nothing else gets to run and other critical functions may be compromised."

He grabbed a pen and made a list as Ravi watched. "Use the pin wiggling to verify that the ISR overruns periodically - you want to make sure there is nothing else offending in there besides that sensor function.

"Next, you gotta do something about that sensor function. Break it into smaller time pieces or figure out exactly what has to be in the ISR. Ideally, get it out of the ISR altogether."

Ravi nodded.

"Finally, use the pin wiggling again to verify that your change makes the ISR happy. Measure its duty cycle; it should be pretty low or we have other problems."

"What should it be - is there a good number?" Ravi asked.

"As low as possible so other functions have enough time to run. Less than 25% is better, but each device is different. From what I see in this ISR, it should be pretty low because there isn't much going on other than updating some timing and motor parameters."

As Oscar rounded the corner out of the lab, he caught sight of Randy at the end of the hall and scooted into a cross aisle to avoid him. He was horribly late on a presentation he was supposed to be preparing for Randy, but he hadn't expected to be bouncing between Ravi and Josie with their debugger-challenged projects.

Randy continued to ride him about late deliverables, but he kept dragging Oscar to meetings where he had to sit, bored, until it was his turn to provide useless status information. Screw the presentation, he decided. Wasn't he delegating projects and mentoring engineers just like Randy wanted him to?

After Oscar left, Ravi thought about the pin wiggling idea and decided it was a neat way to peek into the processor. He set about measuring the total execution time of the interrupt service routine but quickly ran into trouble. The ISR wasn't like a normal function where he could put the pin-wiggling port calls before and after it. It was configured as a periodic timer, which meant that the operating system took care of it. Resting his chin in his palm, he stared at the scope, which still displayed the last captured pulse train.

What he needed was to measure time.

Then he berated himself for being stupid; he may not be able to measure the total execution time including the function call itself, but he could place the pin-wiggling commands just inside the ISR function and still capture a majority of the processing. That way, he would only miss the equivalent of a few assembly-language instructions as the ISR was invoked, which was probably nothing to worry about.

Back in the code, Ravi instrumented the ISR to set the port pin commands at the very beginning and end of the ISR, encapsulating the entire contents of the routine within the two port pin commands. Confidently, he launched the compiler and restarted the program, waiting impatiently to capture the big pause.

To his surprise, the clock rolled over normally.

What was happening? Then he realized that the sensor algorithm function was still commented out. Annoyed, he restored the software to its original buggy state and reran the test. As he had hoped, the scope displayed a series of short pulses followed by a long pulse and more short pulses (Figure 5-5, top). Pausing the scope, he scrolled backwards in time to center the scope display on one of the longer pulses, and then zoomed in (Figure 5-5, bottom).

> **Reader Instructions:** The pulse train Ravi has captured shows when the interrupt service routine is running. The duty cycle is the percentage of time spent in the ISR. From these plots, what is the duty cycle of the ISR for the normal and abnormal pulses? Can you guess why the long pulse is followed almost immediately by another interrupt pulse?

For every 30 pulses in the one-second interval, most were narrow and one was much wider. That meant that most of the time the ISR didn't run very long, but every now and then, it just sat there stuck for a long time before it finished processing.

Finally, he was getting somewhere.

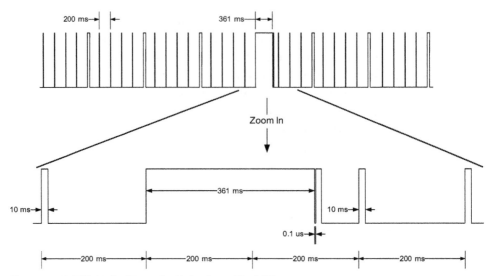

Figure 5-5 ISR Pulse Train for RoboGym Clock Timer.

First, he wanted to verify that the ISR was being invoked at the correct frequency. It probably was, but he figured Oscar would ask if he had verified the obvious. A quick check showed that the rising edges of each pulse were 200 milliseconds apart from one another.

Good.

Avoiding the big fat pulse for the moment, he measured the pulse width of the normal pulses and found that the ISR execution time was usually about 10 milliseconds. In contrast, the big pulse was about 360 milliseconds long. That's certainly an anomaly, he thought, although he was confident that the delay he observed in the clock display was probably about that much.

He wrote the duty cycles on his notepad: The ISR was configured to fire every 200 milliseconds, so a 10-millisecond execution time corresponded to a 5% duty cycle (10 ms / 200 ms). Pretty good, according to what Oscar had recommended. On the other hand, the duty cycle jumped to 180% - 360 milliseconds divided by 200 milliseconds - while the ISR chewed through the multiply and divide instructions in the sensor parameter algorithm. That was certainly not good - the ISR ran way to long. Ravi slid a memory stick into the scope and saved the display of ISR pulses to a file so he could print it out later for Oscar.

Now he had to get that function out of the ISR, but it still had to be executed once a second. Nothing in the routine looked like it really had to be in the ISR, except maybe the **update_hardware_values()** function. Hard to tell.

He could split the function into several parts and execute one part at a time in consecutive ISR calls. He jotted on his notepad:

ISR #1	0 msec	Calculate "sensor_coefficients"
ISR #2	200 msec	Calculate "sensor_sensitivities"
ISR #3	400 msec	update_hardware_values

This would take three consecutive calls, but the functionality would still be completed within one second. He set about measuring the execution time of each but was disappointed to find that coefficient calculations took 110 milliseconds, the sensitivity calculations 170 milleseconds, and the hardware update only 10 milliseconds. This idea wasn't great, unless he split each of the functions into smaller parts. It could still be done within one second, but the ISR code would look, as Oscar liked to say, like a train wreck.

He sighed and tried to think of a different method. Oscar had advised getting the whole thing out of the ISR. Was that possible, while still making sure that the function executed every second?

With a flash of insight, he figured out how to remove all except for a single assignment statement.

Reader Instructions: What is Ravi's idea? Will it work? What will Oscar say?

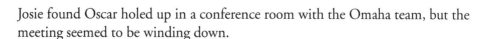

Josie found Oscar holed up in a conference room with the Omaha team, but the meeting seemed to be winding down.

She was too late.

A short time later, he leaned into her cube expectantly.

"Survey says?"

"Yeah, yeah, 100 people surveyed and only one of them has ever heard of auditory debugging. But," she conceded, "he used it to beep when a function was entered so he could listen to the rhythm of the code. Pretty cool, actually."

She handed him a plot (Figure 5-6). "I played around with the auditory debugging method to estimate the Kelly motor position over several box cycles. While I can't verify this with actual numbers, I believe this is a reasonable drawing of the carriage distance over time."

Oscar inspected the plot as she continued.

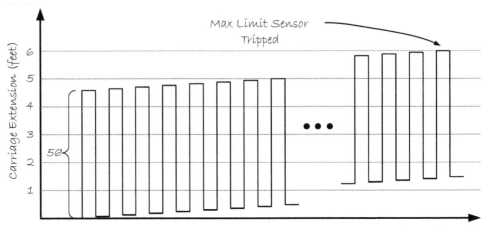

Figure 5-6 The Kelly Labeling Machine's Creeping Carriage Position.

"I changed the main loop so that the motor travels between 0 and 56 inches, absolute position rather than relative position. That will stop position errors from accumulating cycle to cycle."

"Good," Oscar said. "Missed pulses can also be caused if the controller receives a burst of high-frequency pulses. Stiction is another gotcha; if the motor is moving a shaft into something or against something, it has to overcome static friction between the surfaces before it can move."

"What would you do in those cases?" she asked. "Just use encoder feedback?"

"Yes, and you can also use a missing-pulse detector circuit that can trigger the microprocessor directly."

He waved the plot in the air. "But your plot isn't conclusive. You need hard numbers."

"Okay, then, how about this? I checked the weight of the boxes, and many of them are over 600 pounds. The motors aren't rated for that. So I asked Don if we could run some production of lighter boxes, and you know what? All of the sound pitches were exactly the same."

"You're getting warmer . . ."

But Josie had saved the best evidence for last.

"One of the labels is applied to the top of each box. Looking back through earlier boxes, I measured where the label was applied and found it was slightly different after each heavy box. They were offset more and more from the top edge until the failure. After the fix, all the labels are placed at exactly the same spot. No offsets."

To her satisfaction, Oscar nodded without reservation.

"That's conclusive enough for me. They gotta lighten the boxes or order bigger motors."

"You soldered right to the microprocessor chip? How can you see those little tiny pins?" Li Mei looked incredulously at the board, turning it this way and that to peer at the connections.

"It works just right. Oscar showed me how to do it. And no, there are no solder bridges." Ravi boasted about adding wires to wiggle more pins on the processor, and showed her the code. "You just toggle the port pin HI or LO, and record the signal on the scope." He handed her printouts of his experiments and explained how he had measured function execution time without using a debugger.

Li Mei nodded. "This is very different from setting breakpoints and subtracting the timestamps, but can you measure accurately?"

"When a system has an error at nearly 200% percent, fine accuracy is not required."

Li Mei and Ravi both jumped at the sound of Oscar's voice, and turned to face him.

"You can get pretty good accuracy and resolution from a digital storage oscilloscope with a large trace buffer, but this device had such a glaring train wreck in the ISR that we could have debugged it with an old analog scope."

Oscar pulled up a stool to join them. "So, Ravi, I got your email. Amaze me."

Ravi silently prayed he had thought of everything, and began with his abandoned attempt to split the **sensor_params()** function into multiple pieces.

"That was too messy," he explained, "so I moved the call to the **sensor_params()** calculations out of the ISR and into the main routine. Instead of calling that function directly, the ISR just sets a flag. The main routine runs a big endless loop, and it checks the flag and processes the sensor parameters when the flag is set, and then it clears it when it's finished." Ravi handed Oscar a printout of the changed software (Figure 5-7).

> **Reader Instructions:** Before continuing, think about what kind of problems this change might cause. Under many types of conditions, it will not work correctly. Ravi thinks it does - how do you think he verified it?

Oscar quickly perused the code. "Did the clock display really keep accurate time with the bug in it?"

```
unsigned char counter_200_msec;
unsigned char seconds, minutes, hours;
unsigned char LEDControl_f, Recalculate_sensor_params_f = NO;

/* ------------------------------------------------------------------
Name:      TIMER_ISR()
Function: Interrupt service routine to control timing of some system events.
Actions:  This ISR is configured to run every 200 milliseconds.
------------------------------------------------------------------ */

void TIMER_ISR(void)
{
    /* ------- 200 msec ----------- */

    counter_200_msec++;
    if (LEDControl_f == LED_ENABLED)               /* Logical A    */
        update_LED_state();
    update_sensor_control();

     /* ------- 1000 msec ----------- */

    if (counter_200_msec == 5)                     /* Logical B    */
    {
        counter_200_msec = 0;
        seconds++;

        if (seconds == 60)                         /* Logical C    */
        {
            seconds = 0;
            minutes++;
            Recalculate_sensor_params_f = YES;     /* NEW */

            if (minutes == 60)                     /* Logical D    */
            {
                minutes = 0;
                hours++;
            }
        }
        updateClockDisplay();
        update_motor_profile();
    }
}

main()
{
    while (1)
    {
...
        /* New code to compute sensor parameters outside of TIMER_ISR() */
        if (Recalculate_sensor_params_f == YES)
        {
            sensor_params();
            Recalculate_sensor_params_f = NO;
        }
...
    }
}
```

Figure 5-7 Final Software Listing for RoboGym Clock Timer.

With little preamble, the grilling had started.

"I believe so. I measured elapsed time for an hour using an independent timer and found it would be less than a second over an hour's time at most. It wasn't an accumulation of all those half-second pauses."

"Hmm." Oscar mused. "Interesting. What's the time delay between setting the flag in the ISR and updating the parameters in **main()**?"

"Between 4 to 55 milliseconds." Ravi had two concerns about his proposed fix, and Oscar had immediately pounced on the first. Despite a raised eyebrow, Oscar did not reply.

Li Mei slowly leaned back against the bench to observe.

"What was the initial duty cycle of the ISR when it was failing?"

Ah, an easy one. He handed Oscar a printout of his first experiment (Figure 5-5). "The duty cycle of most ISR calls was 5%, taking about 10 milliseconds to execute. Every 30th call was 360 milliseconds, for a duty cycle of 180%." He felt compelled to ask how Oscar had known ahead of time that the error was almost 200%.

"Be careful," Oscar admonished. "The error isn't 180%, but it's close. Remember that ten of those milliseconds are due to legitimate processing." He hoisted an ankle over one knee as he considered the scope traces. "But yes, the error is still appalling."

"Did you see this, Li Mei? This is an excellent example of interrupt overrun, in a big way." Li Mei nodded and accepted a copy of the plot.

"Program control is stuck in the ISR and the next interrupt is trying like mad to happen. I'll show you why there was no cumulative timing error." Oscar pointed to the bottom trace where Ravi had zoomed in on the long pulses. "Right after the long ISR pulse finishes, another pulse happens almost immediately because another interrupt is pending. In the RoboGym architecture, interrupts must be disabled while the TIMER ISR is running so it can't interrupt itself. In addition, the interrupt source appears to be latched - this means an interrupt signal that fired in the middle of the long pulse would still be available for resampling after the ISR completed; otherwise, that missed interrupt would be lost rather than tagged on the end of the long one." He allowed himself a small chuckle. "If the interrupt source wasn't latched, every 30th ISR would be swallowed and anyone using this exercise equipment would be working out about 4% longer than they intended to."

With a final nod, he dismissed the trace on the bench and turned back to Ravi, sitting nervously on his stool.

"What's the new duty cycle for the ISR?"

"Most are 5% like before. Every fifth is now about 70 milliseconds."

Oscar returned to the code listing. "Now here's a trick question."

Ravi groaned internally and waited for the bomb.

"We measured earlier that it took approximately 290 milliseconds to execute the `sensor_params()` functionality before. How long does it take now?"

Ravi inhaled sharply; he had predicted Oscar might go down this line of questioning. Would he be able to surprise Oscar with the answer?

"Because the **sensor_params()** function is no longer processed at the high priority of an interrupt, it takes a little longer than 290 milliseconds to execute. It starts running, but is interrupted once by the ISR before it has a chance to finish. Therefore, it now takes 290 milliseconds plus 10 milliseconds to finish, for a total of 300 milliseconds."

Oscar's face was impassive.

"Prove it."

With a flourish, Ravi placed the second scope trace (Figure 5-8) in Oscar's open hand and announced, "I measured the ISR and the **sensor_params()** execution times simultaneously. The ISR is fixed, and **sensor_params()** runs a logical 10 milliseconds longer than before. Q.E.D." A grin stole onto his face and he turned to see Li Mei smiling back at him.

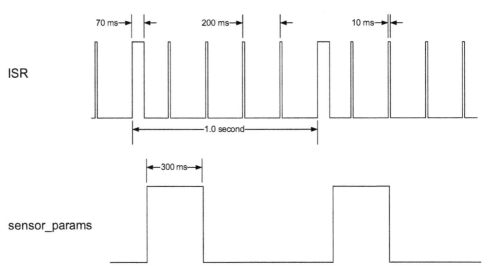

Figure 5-8 New RoboGym ISR Pulse Train.

"What happens if an interrupt occurs between the hardware update for sensor 2 and sensor 3 in **sensor_params()**?" Oscar's monotone interrogation continued with no loss of momentum.

Thrown off guard, Ravi was slow to change gears. "Ah, well," he said, looking back to the software, "I am not sure." The admission was difficult. Was Oscar trying to make him look bad? Li Mei's face was impassive. They sat in silence as Oscar stared blankly across the lab, absorbed in thought.

Shortly, Oscar was back. "I have two concerns. First, removing a function from an ISR means that that function can now be interrupted. You didn't check if anything in the function needs to be protected from interruptions. I suspect that the hardware update commands should be grouped together."

He pulled a pencil from his pocket and showed Ravi how to disable interrupts before the hardware update and reenable them after. "Just place the operations you want to protect within calls to **interrupt_disable()** and **interrupt_enable()**. But turn off interrupts sparingly, only when vitally needed. It builds up latencies in the system and creates other nasty problems." He handed Ravi the sheet.

"Next, do you know how long it takes to execute that hardware update? This is important, because you know what happens if interrupts are off for too long."

Ravi expelled a breath with chagrin; by now he was well aware of what delayed interrupts could do. Luckily, he had the answer Oscar was looking for. "It takes about 10 milliseconds. That should be okay, right?"

Oscar nodded. "Encapsulate those commands to fix that. Finally, you need to check with the hardware folks to see if delaying the updates on the hardware values will be a problem. There will be some jitter in the system now when the **sensor_params()** are applied because the main routine may be off doing other things when the flag is set in the ISR. Just make sure that small delay is acceptable. Also make sure that taking 300 milliseconds to perform the math is acceptable. If it's not, you have some code optimizations to do."

Ravi agreed to check. "How come you weren't surprised or happy when I showed you the scope trace with both pins wigging at the same time?"

"Because I already knew what you did." Oscar uncrossed his legs to stand up and stretch. "When you answered my very first question about the time delay between the interrupts and the function call, I figured you had soldered on another wire.

"I am glad you figured out how to verify your fix. It means we didn't have to sit here all afternoon. I have to call Josie and see if her machine can keep a tune, and then we can all head over to Molly's."

He heard Li Mei giggle and turned to face her. "Perhaps I'll charge you an adult beverage for letting you sit here all afternoon watching other developers work hard."

For a heartbeat, her face registered a stunned expression, but she quickly saw through his threat.

Leaning in conspiratorially he added, "And don't think I missed you taking notes in your little book there, Li Mei, so jot this down too. 'One. Characterize your interrupt service routines. Include only what is critical. If some operation is atomic, you can carefully and momentarily turn off interrupts to complete the operation, but characterize that as well. Two. Low-tech methods like pin wiggling may not be glamorous, but can be mightily effective if used properly. Three. Always reserve a couple of output pins just for software debugging.'"

As Li Mei transcribed, Ravi wondered what to do if the time-critical processing couldn't be removed. He posed the question to Oscar.

"That's when a more thorough characterization can help you." He glanced at his watch, and sat back down with the listing beside him.

"Let's go back to the ISR for a minute. First, look at the four if-statements and figure out how many different ways the ISR can execute. With these four logicals, the ISR can execute eight different ways."

"Eight? Wouldn't that be 2-to-the-4, or 16 permutations?" Ravi asked.

"Normally, but remember that some of the logicals are not completely independent of one another. For example, you don't increment **minutes** unless **seconds** just got incremented." Oscar went to the whiteboard (Figure 5-9). "Let's make a truth table of what code gets executed each pass through the interrupt. Look back at the code (Figure 5-7) - we'll use the letters in the code comments and refer to the four logical code chunks A, B, C, and D. There are eight execution scenarios. See how C and D are dependent upon B being true?"

Possible ISR Paths	If Logical A	If Logical B	If Logical C	If Logical D
1	0	0	N A	N A
2	0	1	0	N A
3	0	1	1	0
4	0	1	1	1
5	1	0	N A	N A
6	1	1	0	N A
7	1	1	1	0
8	1	1	1	1

Figure 5-9 RoboGym ISR Function Processing Truth Table.

Ravi realized Oscar was right. Only the first logical was independent. He also realized that the ISR had not just two but up to eight different execution times, depending on which logicals were true. With a flash of understanding, he realized what to do next.

Grabbing a marker, he joined Oscar at the board and created a second table with function names down the left side and each of the logicals across the top. Li Mei, not missing a thing, kept her pencil poised for additional notes.

"I am putting an X to show which functions are executed in each of the logicals," Ravi said. "I'm not including simple assignments; they don't take long and there are only a few of them."

Oscar nodded in approval.

"I'm also including a column just for the ISR, because one of the functions is executed all the time."

Next, he checked his notes and added another column, labeling it "Execution Time (ms)." Since he had already measured **sensor_params()**, he entered '290' for that function. He also knew that the fastest ISR execution time was 10 milliseconds, so he assumed that was the execution time when no logicals were true.

Li Mei asked, "Ravi, from your table, it looks like the biggest problem occurs when A, B, and C are true. That's when all the functions run, right?"

"Yes, that's true. And now I realize why some of the ISR execution times were still a little big after I fixed the software. I never looked at it this way before." Moving to his computer he added, "I just want to measure the time to execute the other three functions so I can fill in this table." Shortly, he had recompiled, checked the scope, and completed the table (Figure 5-10).

Function	ISR	Logical A	Logical B	Logical C	Logical D	Execution Time (ms)	Execution Time (ms)
Frequency of Execution	200 ms	random	every second	every minute	every hour	Old	New
update sensor control()	x					10	10
update LED state()		x				1	1
updateClockDisplay()			x			30	30
update motor profile()			x			30	30
sensor params()				x		290	
						361	71

Figure 5-10 RoboGym ISR Function Execution Times.

"Adding up the execution times, this shows the original worst-case scenario at about 361 milliseconds. After I changed the code, the worst case is only 71 milliseconds." He turned to Oscar. "The math works - it all adds up!"

"I think you've got it, Ravi. This is a straightforward way to see where the cycles are being used in the ISR. If you go back to the truth table, you will see there are only four possible execution times because there's nothing much left in C and D now."

"Yes, I see." Ravi was hunched over his pad, adding numbers. "The new possible ISR execution times are 10 milliseconds, 11 milliseconds, 70 milliseconds, and 71 milliseconds."

Oscar crossed his arms and thought for a moment. "Looking at your table, now let's say that the elapsed time clock and the motor functions were each 100 milliseconds instead of 30. Then the new ISR would run for 211 milliseconds. Assuming that all that functionality had to remain in the ISR, what would you do? The ISR shouldn't overrun its 200-millisecond interval, and it needs to let other software run as well."

As Ravi paused to think, Li Mei jumped from her stool to grab the software listing.

"I think I know. I would add another if-statement to check when the `counter_200_msec` variable reaches 4 instead of 5, and then move some functions in there." Li Mei said, looking back and forth between them. "If I check with hardware guys that `update_motor_profile()` can be executed 200 milliseconds sooner, then the worst case is now 111 milliseconds two times every second, and the other three ISRs every second are just 11 milliseconds."

"Yes; that would distribute the time more effectively without overrunning the interrupt." Oscar pushed away from the bench and brushed off his hands. "Looks like you two have it all figured out - no reason for me to be here." Ravi's face flushed at the unexpected praise, and he shared a smile with Li Mei.

"Li Mei, what the heck is that drink?" Ravi leaned over to look in the tall curvaceous glass with skewered pineapple and cherry.

"I asked my friend for a drink with some history, like that black beer Oscar likes. She told me to try a Singapore Sling." Taking a sip she added, "It is quite good."

"Ah, that's a foofoo drink." Oscar waved a hand to dismiss it. "Good Irish stout is the optimal beverage to contemplate embedded development, which is why we are again comfortably ensconced in the most comfortable corner booth at Molly's." He drained the rest of his pint in emphasis.

Josie shot him a dirty look. "Oscar, the only thing Irish about you is your username. 'oshelley' hardly qualifies you as a good Irishman."

Oscar changed the subject. "So you guys had to do some nontraditional debugging lately, huh? Wiggling pins and whistling tunes?"

Ravi and Josie both nodded, smiles tinged with relief. Josie sagged against the seat. "That was frustrating and amazing at the same time. I thought you were yanking my chain, telling me to make the machine sing. Do you have any idea what Don said when he heard beeps and chirps coming out of that machine? He thought I was nuts!"

"Don't worry about what people think; just worry about getting the problem fixed." Oscar leaned in. "I read some interesting research on auralization. Musical phrases, chords, and complex tones can be linked to program language constructs like if-statements and loops in order to understand program behavior. They've tested that novice debuggers can detect defects in program flow just by listening to the program. We used simple pitch intervals to detect when the relative carriage movement changed. Pretty powerful, eh? [2]

"So, Josie, what debugging tips can you add to Li Mei's List?"

Josie looked at him in surprise, but saw Ravi and Li Mei nodding.

"He found out about my list," Li Mei admitted. Looking at Oscar, she added conspiratorially, "And if he is not nice, I will not share my wonderful list when he has a big debugging emergency next time." Josie choked back a laugh and marveled at Li Mei's humor and presence. She was fitting in almost *too* well!

Josie dipped a nacho and composed herself. "Okay, let me think what pearls of wisdom I can add."

Additions to Li Mei's List of Debugging Secrets

Specific Symptoms and Bugs

- Use absolute (rather than relative) motor position references for regular, repeating activities. Send the motor home regularly.
- Don't assume the motor does what you tell it. Check it.
- Make sure only time-critical functions are included in Interrupt Service Routines. Verify the duty cycle. Split functions or use flags to remove noncritical code.

General Guidelines

- The sense of hearing can discern fine differences in tone and rhythm, and can also be used as a heartbeat or a code coverage flag.
- Simple debugging methods like pin-wiggling can be powerful and unobtrusive. Think about exactly what information you would like to get from the embedded system, and choose the best tool accordingly.
- Be clever. Don't worry about what other people think. Just worry about getting it fixed.

Chapter Summary: The Case of Two Inaccessible Microprocessors

Ravi and Josie must debug their projects without the luxury of a debugger or emulator. They must identify exactly what needs to be measured and develop outside-the-box alternative methods. Ravi has no debugger for the RoboGym module. He measures execution time of various functions in an interrupt service routine by toggling a spare port pin and viewing the signal on an oscilloscope. Similarly, Josie can't touch the Kelly Manufacturing labeling machine for safety reasons. She measures distances using her sense of sound to hear tones that correspond to distance traveled.

The Case of the Creeping Slider Arm (Difficulty Level: Moderate)

The Problem Symptom(s):
- A box processing machine at Kelly Manufacturing stopped after running for hours. After a power cycle, it ran for several more hours before stopping again.
- Sometimes the machine would run for months without error.

After each failure, three things were noticed.
- A red indicator light in the control panel was illuminated.
- The motor carriage was extended many feet, blocking the assembly line.
- The rotating foot on the end of the carriage was not in position to pull in a box.

Targeted Search:
- The lit indicator was connected to a magnetic proximity sensor on the motor carriage.
- In software, the proximity limit sensors were handled by operating system software that is not visible for debugging.
- The code that waited for the carriage to fully extend didn't contain errors.

The Smoking Gun:
The carriage extended the commanded distance but retracted less than the commanded distance.

The Bug:
- Relative distance (motor counts) were commanded rather than absolute distances. On retraction, the stepper motor could not pull the over-weight boxes up the slight incline and motor pulses were missed.

The Debugging Method Used:
- Coding actual motor position as a tone, and then using auditory debugging to hear changes that correspond to different stopping positions each time a box was processed.

The Fix:
- Changed motor commands from relative to absolute to prevent cumulative position error.
- Changed command to retract the carriage 56 inches into a command for the carriage to return to the home position. ("Home" is commonly set during initialization to establish known positions for the hardware. On error or for cyclic behavior, motors are often sent "Home.")
- Added software to compare the box weight to the maximum limit and warn operator when boxes are too heavy. Customer advised to upgrade motors for heavy boxes.

Verifying the Fix:
- Qualitatively verified new carriage arm position using the auditory method.
- Inspected the label positions on light and heavy boxes to indirectly verify that the error for heavy boxes did not accumulate.
- Verified that overweight boxes generated a warning message.

Lessons Learned:
- A device can be debugged without touching it.
- Motor positioning should be verified.

The Case of the Hesitating Clock (Difficulty Level: Moderate)

The Problem Symptom(s):
An elapsed-time clock feature on the RoboGym control panel took too long to roll from 59 to 00 seconds. The symptom was periodic: once every minute.

Targeted Search:
- Many references to clock updates appeared in the ISR.
- Symptoms suggested that processing caused a delay right before the clock update. Functions in the ISR were identified as suspects.

The Smoking Gun:
- Time to service the ISR for the 59-to-00 transition was much longer than the normal duty cycle, almost twice as long as the normal ISR period itself.

The Bug:
The ISR contained a math-intensive function that caused it to exceed a 100% duty cycle once every second.

The Debugging Method Used:
- Pin wiggling to characterize the duty cycle of the ISR by monitoring a spare output port pin with an oscilloscope.
- Patiently watching the system's behavior.

The Fix:
- Replaced the entire function call in the ISR with a flag since the math performed in the ISR was not critical.
- Added an if-statement to main() to check the flag and call the function.
- To protect the atomic hardware calls, disabled interrupts just long enough to update the hardware configuration.

Verifying the Fix:
- Measured the new duty cycle of the ISR to be 5.2% (10.4 msec), with periodic jumps to 35.5% (70 or 71 msec) on 1-second rollovers.
- Verified that the moved hardware function also completed in a timely manner.

Lessons Learned:
- Choose the right tool for the job.
- Choose robust and objective methods to characterize symptoms so the same method can be used to verify the fix.

Code Review:
- The Hesitating RoboGym Clock software was reasonably well written and is self-documenting with descriptive names and consistent brackets and spacing.
- The Creeping Slider Arm software contains useful comments to compensate for the cryptic language commands. Additional spacing between functional chunks would increase readability.

What Caused the Real-World Bug? Within the Patriot Missile's guidance system, an internal clock, accurate to 0.1 seconds, helped discriminate the flight path of incoming threats to separate friendly F-16s from an evil dictator's last-gasp missile launch. But the number "0.1" cannot be represented accurately in binary using the Patriot's 24-bit nonfloating point word. Over time, errors accumulated. In the 100 hours this Patriot battery had been running, the error totaled 0.34 seconds, which corresponded to a predicted Scud flight-path error of more than half a kilometer, enough to cause the code to figure the Scud was really a friendly.

The irony: the problem was known and had already been corrected. Fixed code was enroute and arrived at the site the next day. [3]

References

[1] Ganssle, J. (1992), *The Art of Programming Embedded Systems*, Academic Press.

[2] Vickers, P. and Alty, J.L. (2002), "Using Music to Communicate Computing Information," *Interacting with Computers*, 14 (5), 435-456.

[3] GAO/IMTEC-92-26 Patriot Missile Software Problem, United States General Accounting Office, Information Management and Technology Division, B-247094, February 4, 1992, accessed from http://www.fas.org/spp/starwars/gao/im92026.htm.

Additional Reading

Baecker, R., DiGiano, C., and Marcus, A. (1997), "Software Visualization for Debugging," *Communications of the ACM*, 40(4):44–54.

Ganssle, J. (2004), *The Firmware Handbook*, Burlington, MA: Elsevier (Newnes imprint).

Parker Hannifin Corporation (1998), *6K Series Command Reference*, Parker Automation, p/n 88-017136-01 A.

Vickers, P. and Alty, J.L. (2002), "When Bugs Sing," *Interacting with Computers*, 14 (6). 793–819.

Chapter **6**

If I Only Changed the Software, Why Is the Phone on Fire?

"Please leave a message . . . BEEP."

"Oscar? It's Randy. We got a big problem. Mike called from the trade show floor. They got the phones we shipped overnight and apparently one of them caught fire when he was doing an Austin Monitor demo for a customer. I don't care how late everyone was in the lab last night. You've got to get in here now."

Listening to his voice messages as he pulled out of the computer megastore parking lot on the way to work, Oscar nearly T-boned a passing car.

He pulled over and wiped the hot coffee off his dashboard, and then constructed a terse text message. After a short debate with himself, he selected a different recipient and hit SEND.

By the time he reached Hudson Technologies, it was well after his normal early-morning arrival time. The lab was still empty, and he surveyed the damage from the last few weeks of late nights. Empty bottles and snack wrappers littered the benches and the whiteboard was filled with cryptic variables and numbers. Last-minute changes and bug fixes hadn't been fully tested before final phone and monitor samples were shipped overnight using the carrier with the latest possible pick-up time. Testing had continued afterwards with the fervent hopes no other problems would be found.

But something had gotten past them.

———————————

Ravi stood barefoot and bleary-eyed, staring at the words on his mobile phone, mortified that he'd done something to break the phone right before the biggest trade show of the year. He yanked on a pair of jeans and pulled a fresh shirt from the dresser.

No doubt, Oscar was going to kill him.

———————————

Josie stared out the window, straining to see squares of farmland through the clouds below her. She turned the worry stone over in her hand and wondered if this would be her last trip to visit her grandmother. The string of late nights at work and her early rise to get to the airport soon took their toll, and she was asleep.

———————————

Back in the lab, Oscar tipped back on the stool and stared at the ceiling tiles high over the benches, mindlessly flipping his badge around the lanyard. He was only marginally aware of Li Mei sitting at her bench beside him, hovered intently over a stack of documentation he'd scraped together for her. Josie was the lead for this project but she was unreachable and Ravi was still nowhere to be seen.

Patience, he counseled himself.

While he waited, he thought back over the myriad projects he'd worked on since graduating with a masters in electrical engineering, back in the days before the internet had gained mass acceptance and most of the computer number crunching was still done on UNIX mainframes or using PC-based command line programs. Way before web pages and internet commerce and icon-based interfaces, but certainly post-punch card. He'd written his first embedded programs on a Hitachi 6305 microcontroller, first coding the LCD display driver so he could use the 4-digit display for debugging. Then he'd implemented the LED driver and pushbutton code to interface with the real world.

Before that fateful experience, his exposure to programming had been the typical "Hello World" and sorting algorithms, stuff he tagged boring and not particularly relevant for an electrical engineer. But the ability to toggle individual bits in software and watch motors spin or displays change color - now *that* was cool.

He was hooked.

Going to class took a back seat to crunching out new gadgets on the soldering-iron-scorched coffee table in his dorm room. Over the years, simple interfaces gave way to more complex control of electrical systems, with real-time signal processing, support of multiple serial ports, real-time operating systems, and wireless communications.

Now, sometimes he had to stand back and smile, watching a powerhouse microprocessor happily crunch a million lines of code, communicating in the Gigahertz spectrum, all while running off a 3V battery.

Amazing how things had changed.

The sound of Ravi's footsteps smacking on the vinyl floor jarred him back to reality, and he began without preamble.

"I asked Mike to email me the sequence of events that led to the fire. As close as he can figure, this is it."

As he read, he also bulleted the contents of the email on the freshly cleaned whiteboard. "Mike writes, '1: Turned on phone. 2: Waited for phone to acquire a system. 3: Turned on the Austin Monitor. 4: Turned on the pulse oximetry sensor and collected some data. 5: Launched the Austin application on the phone and waited for it to make connection with the Austin Monitor. 6: Made a phone call to upload the pulse ox data, and set the phone in the cradle.'"

Looking to see that he'd transcribed the items correctly, he returned to the email. "Mike goes on to say, 'I was talking to the customer during the data transfer, and then we smelled smoke and saw a small flame from the side. I pulled it out of the cradle and yanked the battery. It was hot at the bottom, but not up near the display.'"

He dropped the email on the bench and asked, "What the heck could have happened?"

"I tested my changes really well; full code coverage, Oscar," Ravi implored. He had been fidgeting in his seat since he'd arrived. "I really don't think this is anything I did last night."

"Not now." Oscar silenced him impatiently with a motion of his hand. "I want a list of everything we changed, and everything that could cause a phone to catch fire, and then we'll get Mike back on a conference call."

———————————————————•

"Hi, Mike. I've got Randy, Ravi and Li Mei with me."

"Oh, Josie's not there?" Mike sounded disappointed.

"She's away, but Li Mei knows a little about the Austin Monitor so she'll fill in."

After quick pleasantries, they huddled near the speakerphone and listened to Mike vent over the background din of the Worldwide Telecom Show.

"Oscar, I don't have to tell you this, but we have to demo this feature at the show. Right now we are telling everyone that we have a bad pulse-ox sensor but that's only good for maybe a few hours. We're showing a working demo on video, but it just doesn't have the same impact."

"I know. What did the guys from Omaha Telecom say?"

"They've never see this happen before, and they're pointing the finger at us and Austin."

"Figures." Oscar flattened a paper on the spotless mahogany table. "Anyway, some questions for you. From your email, after you launched the Austin app on the phone, did it actually communicate with the Austin Monitor?"

"Yes, I saw the 'Connection Established' screen."

"Okay, so those two are communicating with one another. When the phone tried to upload the data to the server, did it finish sending?"

"I don't think so." Mike paused. "It displayed the 'Connecting' screen, but it never beeped like it does when it's finished. But I was talking to the customer and not watching the screen closely."

"Did you test it before showing it to anyone? Anything else suspicious?"

"Absolutely I tested it." Mike was emphatic. "No one was on the show floor yet. I turned everything on, put the sensor on my finger, collected data, uploaded the data, and checked it on the server. It took a little while to acquire service, but I could make phone calls. All the test data I uploaded should be on the server. You can check it."

"Yesterday was crazy with the connection problems on our end," Oscar grumbled, "but at least you got the Monitor to make connection with the phone. Check the phones - what version of software is in there?"

While he waited for Mike's answer, Oscar leaned back in the substantial leather conference room chair and stared at the ceiling. If they'd overnighted the wrong phones . . .

"The software version is R12.23_06.16."

Ravi confirmed the version and then read back a list of the serial numbers for the phones they had hurriedly shipped. They matched.

"Just checking the obvious," Oscar clarified before continuing. "So, you mentioned a phone cradle. Is that the one that's also a battery charger?"

"Yes, and it was plugged in."

Oscar raised an eyebrow slightly before continuing, and noticed that Li Mei had taken it upon herself to capture Mike's responses.

"Did you hear that problem on the news with some cell phones exploding or getting really hot? Most of the explosions were caused by counterfeit or bad batteries. Do you have genuine ones to swap out?"

"All of them came right from the manufacturer, but I have a few extra. I am walking back to the booth to try one now."

Mike paused, then asked quietly, "How do I test this thing without it catching fire again?"

The additional information from Mike was useful, but inconclusive.

After reinitiating a data upload using the Austin phone application that Josie had written, Mike had warily placed the phone back in the cradle. He rested his hand on the side of the phone to track the temperature. While the team listened on the speakerphone, he relayed the connection messages on the phone's display, and then shortly reported that the phone was getting uncomfortably warm to the touch. He dropped his own cell phone to yank the battery pack from the second demo phone before anything got damaged.

So it probably wasn't a bad battery. Oscar was disappointed; he'd immediately thought of the exploding batteries and had hoped a bad battery was the cause, but it seemed they were not to be so lucky.

The last-minute scramble before a trade show in any industry was typical, and this was no exception. Hudson Technologies was developing the Austin Home Medical Monitor to collect real-time data from homebound or elderly individuals and transmit it via cell phone to a central database for long-term monitoring and emergency services. Everything was wireless, from the sensors themselves to the cell-phone link. Josie had written most of the software for the stand-alone monitor and the cell-phone application, with driver and integration help from Oscar and Ravi. Their contact, Dave, from B&L Omaha Telecom had helped verify that the application was properly integrated into the mobile phone software architecture, but the Omaha team had had their own last-minute bug fixes to deal with before the show.

Managing software changes on three devices while keeping all communicating with one another had been a major headache. The Austin phone app and Austin Monitor programs were functional; but barely so. They'd recently optimized the data packet format to increase the effective data bit rate, and the initial connection time had been a problem until Ravi had rewritten part of the data rate negotiation software. Currently, the Austin Monitor only supported the pulse oximetry module although several other sensor types were in the works.

Oscar thought about what to tell Mike. Throughout Oscar's tenure at Hudson Technologies, Mike had been a great resource in marketing, spending time in the lab to understand the real technical issues.

But now Oscar rued letting Josie give him an unofficial demo of Austin after it worked correctly the first time. Even though the software had managed to upload the medical data correctly a month ago, it had crashed the phone shortly after.

Mike, in his excitement over the progress on the project, had mentioned the "milestone" to his management. And while Hudson Technologies had not originally committed to support the Worldwide Telecom show, Omaha Telecom had sweetened the pie if the Austin product with just one sensor could be demonstrated.

Without a follow-up to Mike or engineering, management had accepted Omaha's generous offer.

"Has anyone ever seen anything remotely like this problem during development?" Oscar asked.

Ravi shook his head. "And there is nothing in the defect-tracking system either."

"We need to reproduce this." He looked around the lab. "Li Mei, you don't know this system as well as the rest of us, so you can be the typical unsavvy user. Based on what you just read in the manual, connect everything together and see if you can induce flames."

"And Ravi, go see Tom and ask him what in hardware besides the battery could cause a cell phone to catch fire. I need to call Dave over at Omaha and ask them if they have the tools to load new phone software at the show if we need to."

While Li Mei worked through the Austin software next to him, Oscar thought about what could be different about the conditions in the lab from those on the trade show floor. They'd never seen the problem here, but it had already occurred twice at the trade show with two different phones, using different batteries.

They only had two working phones here, so they'd need to be careful.

The software in all phones was the same. But something was different enough to cause a fire. What could it be?

Even though all phones ran the same software, each had several configuration settings that could change which parts of the software were executed. He scrolled through the phone's menus looking for setup information and found references to entries called PRL, SID and NAM. After a quick search on the internet, he learned that Preferred Roaming List (PRL) contained a list of frequency bands, channels, and System Identification Codes (SIDs) that the phone uses to find service. The Number Assignment Module, or NAM, was the memory containing the phone number and the electronic serial number, or ESN. Maybe Dave could shed further light on the possibility of a configuration issue.

He saw Ravi return to the lab and watched him step up behind Li Mei as she cautiously collected and uploaded data.

"What did Tom say?" Oscar was impatient.

"He asked about the batteries right away, but I told him probably not. He suggested the battery charger and the charger hardware in the phone. Any situation with large changes in battery current like charging, transmitting in low signal-strength areas, especially in analog mode, or playing a lot of video or multimedia applications across the network."

As Ravi relayed Tom's suggestions, Tom came up behind him and took over when Ravi finished. "The current drain during a cell-phone call has peaks well over 1 amp. The backlight for the display uses more juice and the flash on the camera can produce some high current spikes."

"If all that happened at the same time, maybe we'd see flames, but I doubt the software would let that happen."

"Probably not."

"We should also check the antenna connections. If the phone can't acquire a strong signal, something could be wrong in the connector or the antenna layout on the board," Tom added.

"Something else to ask Dave." Oscar drummed the bench with his fingers and called over his shoulder, "Any luck, Li Mei?"

She slid off the stool and joined them, holding a cell phone in each hand. "I have tried each one of these several times. Power-cycling, not power-cycling, collecting and sending large and small data files, and in and out of a charger." With a slight dip of her head, she concluded, "I guess I am not a stupid-enough user to make this fire."

Oscar grunted amid the laughter. "I'll let Mike know who you think the stupid user is, after he helped you with your specs last time."

But the levity quickly dissipated and the mood turned sober again.

"So. Li Mei, what does your list say we do next?"

Oscar noted with some satisfaction the small smile of pride that came over Li Mei's face at the mention of "her" list. He also ignored Ravi's agitated body language in response to the subtle praise.

She retrieved her notebook and consulted it for a short moment, and then handed the system diagram to Oscar (see Figure 6-1). She confirmed her understanding of the data path from the sensors to the Austin Monitor, and then through the mobile phone to the base station, internet, and finally to the repository on a local computer. At Oscar's nod, she began her analysis. "We have interviewed almost everyone who saw it fail. I doubt, however, that we should interview the customer."

"That would be a Good Call," Oscar conceded.

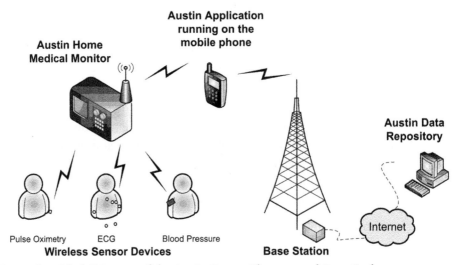

Figure 6-1 Block Diagram of the Austin System Elements and Data Path.

"Also we know the correct system behavior, but we have not been able to reproduce the failure. Even though we haven't seen it, I would classify it as a constant error for a particular set of conditions that we have not yet identified."

She set her notebook back on the bench and added, "I think we should try to understand those different conditions at the show."

"I agree, and I think I know a way to do it," Ravi rushed in. "We used a base station simulator to develop Austin so we wouldn't crash a commercial cell tower. They're not using a simulator at the show. They're live."

Oscar nodded with satisfaction. "Excellent deduction. But recall that this feature *has* been run on the commercial network, but only a few times before the show to make sure it worked.

"So, we will configure our base station to simulate the conditions on the show floor, but first I want to outline our plan of attack once the simulator is set up. Having Li Mei roll the fiery dice all afternoon isn't ideal. So, what's next on the list, Li Mei?"

"We are doing the next item by brainstorming root causes, and asking for input from other groups like hardware and marketing." She hesitated before continuing. "I also have 'Resist the urge to launch the debugger and review the software listings.'"

"Ah, I see Josie's hand in this list!" Oscar resumed flipping his ID badge and affirmed, "That is excellent advice, for where would we look right now? We haven't narrowed down the search at all, and the problem could also be hardware or configuration-based. The software is proprietary, so I doubt we'll be granted access.

"We can't control inputs to the system," he told them. "Users do strange things, press keys at the wrong time, unplug things prematurely, and the software has to handle the illogical inputs in a nice way."

He turned to his computer and opened a file. "I read a good quote the other day. 'Debugging is the process of determining *why* a given set of inputs causes an unacceptable behavior in a program and *what* must be changed to cause this behavior to be acceptable.' [1]

"So this acknowledges that *we* are the ones responsible for handling the unexpected, not the user." He looked between them. "Let's be logical and identify the inputs."

Ravi jumped to the whiteboard to capture their ideas.

Inputs	Culprit?
1 - Wireless link between sensor and Austin Monitor	Low
2 - Wireless link between Austin Monitor and phone	High
3 - Configuration settings for real base station	High
Different time zone	
Different channel frequency and SID	
Real base station versus simulator	
4 - Operator	Medium
5 - Battery charging	Medium
6 - No debugger running	Low

Li Mei looked puzzled. "Ravi, why do you have 'Operator' at a medium probability? Didn't I just prove that the operator is not the problem?"

"Because of everything Tom said. If the phone gets into some high-current modes like multimedia, voice transmission or charging, the operator caused it."

We don't really know for sure what Mike did. But we do know that what you did should be typical for the Austin feature and that didn't reproduce the problem."

"I guess so. And I also didn't have the debugger connected."

"That's why I set it to low," Oscar told her. "You know that having the debugger connected and running affects performance, but once I had a system that ONLY worked with the debugger. Turned out the extra time delay prevented a race condition that only showed up when we thought we were done debugging and ready to ship."

Li Mei cringed. "That must have been nasty to figure out!"

"It was." Oscar didn't elaborate. "And we tried the charger already, so the base station is still an obvious difference but we don't know why."

"Well, one thing is transmit power," Ravi said. "In the lab, we are right next to the base station simulator, so our phone doesn't have to transmit with a very strong signal. Maybe at the trade show the signal is really weak and the phone has to transmit at a higher power level. That drains power and can make the phone warm. We can simulate that problem by configuring our base station to be very weak."

Oscar continued to mull over the list, and then popped his cell phone from the holder and voice-dialed Mike.

"Hey, Mike, Oscar. Listen, how many bars are you getting on the show floor?" He thought Ravi's weak-system theory was a good idea; the phone would have to transmit at a higher power to be heard by a base station it thought was further away.

But Mike came back with different news; the phone had acquired an analog system rather than a digital system.

That was odd. The first wireless networks were analog, and many phones still supported analog for roaming into rural areas where digital hadn't been deployed. But analog could be a power hog and tended to drain the battery faster. The show floor should be spewing enough digital for everyone to acquire a system.

His heart skipped and he felt an adrenaline rush of excitement.

"Analog could be the problem, Mike. Can you get it to acquire a digital system instead? That'll drain less power, less current, maybe no flames."

"That's easy - I can exclude analog systems from the System Information menu. Let me do that now."

From Mike's voice, Oscar could sense recognition of a possible solution, and he saw Li Mei and Ravi sit straighter on their stools as they listened to his side of the conversation.

Several moments later, all three could hear the invective through Oscar's earpiece before Mike reported that the phone again overheated.

"Sorry, Oscar - that was a great idea and probably why the phones were needing to be charged so often. I'll exclude analog on our demo phones, but that doesn't fix the fire problem."

Ravi slumped back down on the stool and turned back to the whiteboard to scan the HIGH and MEDIUM items. He knew the link between Austin and the phone was still suspect, because he'd made changes last night to fix some messaging problems that delayed the connection between the two devices. But he felt adamant that his fix was good and looked elsewhere for the problem.

The Hudson team was only responsible for the Austin items, but he was still suspicious of the charger. The Omaha team said it was working fine. On the other hand, the fire occurred while the phone was in the charger. Tom said charging circuits could cause fires, and he didn't want to let that possibility go ignored.

To his surprise, Oscar actually agreed with him, and he listened as Oscar described the test they wanted Mike to run: logging debugging messages while the phone was charging in the cradle. Ravi had become familiar with the wealth of debugging information that the phone could produce: every message sent, task switching, semaphores, logical decisions and even custom messages that he and Josie had requested to debug the Austin application interface.

Messages he'd used last night to validate his changes to the wireless interface.

But after those short cryptic messages were expanded into descriptive text and data values, the parsed log files could balloon to megabyte text files of information to sift through. Logging also required implementing some invasive code changes.

"Mike is running the test and will upload the log file to the server," Oscar reported. "Would you grab it when it gets there, Ravi?"

"I'm on it." He moved to his bench and downloaded the raw file from the server when it arrived. Oscar and Li Mei joined him as the parser program ran, and together they scrolled through the beginning of the output file. (See Figure 6-2.)

"Josie and I looked at these logs before," Ravi said. "The left column is the operating system task that's running, and the right is the message or action that the phone executes. I can also insert timestamps if we need them, but I don't know most of these tasks. We only dealt with the APPS task when the Austin application ran, and the CALLPROC task that controls the call processing states to set up the data phone call. Also, MS means Mobile Station - that's the phone - and BS is the Base Station."

```
CALLPROC:    MS Idle State                         SEARCHER:    Tune to new frequency
BATT:        Battery update available msg          CALLPROC:    Pilot Acquisition Substate
PILOT:       Pilot Set Maintenance                 SEARCHER:    Pilot acquired!
POWER:       Battery level low                     UI:          Update icons/clock
POWER:       Fast charging rate 100%               CALLPROC:    Sync Channel Acquisition Substate
POWER:       Store charging state                  RX:          Receive Sync channel msg
CALLPROC:    System Determination Substate         CALLPROC:    Timing Change Substate
SEARCHER:    Tune RF/IF                            BATT:        Battery update available msg
CALLPROC:    Pilot Acquisition Substate            CALLPROC:    MS Idle State
SEARCHER:    Pilot acquired!                       RX:          Receive Overhead Msgs
UI:          Update icons/clock                    PILOT:       Pilot Set Maintenance
CALLPROC:    Sync Channel Acquisition Substate     POWER:       Battery level low
RX:          Receive Sync channel msg              UI:          Update icons/clock
CALLPROC:    Timing Change Substate                RX:          Receive Overhead Msg
BATT:        Battery update available msg          CALLPROC:    Update Overhead Substate
CALLPROC:    System Determination Substate         TX:          Registration
SEARCHER:    Tune RF/IF                            RX:          Receive Overhead Msg
CALLPROC:    Pilot Acquisition Substate            CALLPROC:    Idle State
SEARCHER:    Pilot acquired!                       PILOT:       Pilot Set Maintenance
POWER:       Battery level low                     UI:          Update icons/clock
UI:          Update icons/clock                    BATT:        Battery update available msg
CALLPROC:    Sync Channel Acquisition Substate     POWER:       Battery level medium
RX:          Receive Sync channel msg              POWER:       Fast charging rate 75%
CALLPROC:    Timing Change Substate                POWER:       Store charging state
CALLPROC:    MS Idle State
RX:          Receive Overhead Msgs from BS         (Log file continues for several pages)
PILOT:       Pilot Set Maintenance
UI:          Update icons/clock
RX:          Receive Overhead Msg
CALLPROC:    Update Overhead Substate
TX:          Registration
RX:          Receive Overhead Msg                  Key:
CALLPROC:    Idle State                            RF: Radio Frequency
BATT:        Battery update available msg          IF: Intermediate Frequency
PILOT:       Pilot Set Maintenance                 MS: Mobile Station
PILOT:       Idle Handoff                          BS: Base Station
POWER:       Battery level low                     RX: Receive
UI:          Update icons/clock                    TX: Transmit
CALLPROC:    System Determination Substate         UI: User Interface
```

Figure 6-2 Debugging Log File from Idle Phone at the Worldwide Telecom Show.

Reader Instructions: This is a sequential list of activities in the phone. You are interested in charging. What task(s) are related to charging? Does the charging algorithm seem reasonable? Can you figure out what else is happening? (It may be confusing in the beginning unless you're in the telecom biz, but that's okay.)

"Does anyone have a block diagram of the phone?" Oscar asked the group.

Ravi retrieved a sheet of paper and placed it on the table in front of the three of them. Most of the blocks were straightforward, although he didn't understand some of the phone services such as signaling and link layer. (See Figure 6-3.)

Reader Instructions: What parts of the phone are involved in charging? What parts are involved in sending data using the Austin application?

"The logs show a BATT task and a POWER task." He pointed to the page. "It looks like the BATTery task monitors the battery state and sends messages to the

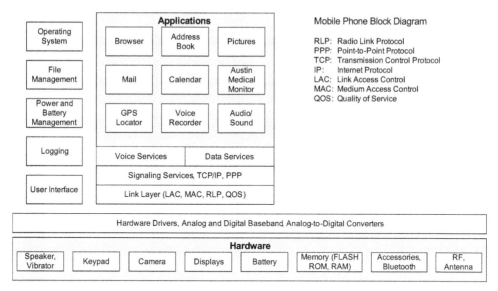

Figure 6-3 Block Diagram of Mobile Phone Functions.

POWER task. The POWER task must evaluate the new battery information and make a decision about the charging rate." He explained that he had found several fast charging rates, such as 100% and 75%, and a lower trickle charge rate, after each BATT task entry. "The battery hardware has temperature and voltage signals that are monitored to decide when the battery needs charging. The battery gets warmer when it charges, so it checks the temperature to reduce the charging rate to prevent any thermal problems."

"We have a thermal problem. A big one," Li Mei said. "Could it be the battery?"

Oscar had been scanning the sheet, and faced her. "This log file shows a nice communication between the BATT and POWER tasks. If the POWER task needs to change the charging rate, it issues another command, like this one." He pointed to the line indicating a fast charging rate of 75%. "If no change is needed, then nothing else is logged."

"Then it looks like the charger is working correctly. Way down near the end of the file it drops to trickle charging. That shouldn't cause a fire." Ravi was fairly confident of his assessment. Seeing Oscar nod in agreement, he was about to recount his battery experience on a different project when Oscar's phone rang. He sighed and put the schematic away.

It was Mike again.

From Oscar's side of the conversation, they could tell that Mike continued to apply pressure about demonstrating Austin and questioned last night's connection problems as the cause of the fire, but something Mike said must have caught Oscar's attention because he suddenly straightened.

"Hold up, Mike. You said the phones are getting warm even when you are not using Austin? How warm?" Oscar paused. "Last time we talked, you said that everyone was having to charge the phones more than usual. Didn't you switch to digital? What's going on?" He grabbed the charger and read the model number to Mike. "Yeah, same ones. I think you guys can use them to charge the phone normally. Just don't use the charger when you are demoing the phone for now - swap fully charged batteries in and out. I'll call you back."

Abruptly, he ended the call.

"Austin isn't the cause of the problem. Something else is wrong, and Austin is just making it worse. Sending it over the edge."

Oscar had begun to pace and talk aloud. Heading back to the whiteboard, he grabbed a marker and drew a black line through Input #1 and #2 and then scanned the rest of the list. Ravi slumped in relief. It was the first sign from Oscar all morning that he believed Ravi's changes from last night had been correct.

"Also remove #5?" Li Mei ventured.

"No, battery charging is still involved since they have to do it so often. But what is *causing* the phone to charge so much? Is it really high battery drain, or is it charging inappropriately?"

After a moment, he answered his own question. "Increased battery drain is more obvious since they aren't having explosions without Austin. But something about Austin makes it worse. What is Austin doing?"

Li Mei ventured, "Josie explained to me that the Austin application in the phone establishes a data call with the base station to upload the medical data packets. The phone also has to talk to the Austin Monitor using Bluetooth at the same time." She broke into a smile. "Isn't that neat? Two different wireless links at the same time? And the Austin Monitor uses a completely different protocol to talk to the individual sensors like the pulse oximetry sensor. That is amazing to me."

Oscar nodded. "I wonder if there's just so much going on in the phone that it's draining power more quickly than the folks at Omaha Telecom planned." He picked up the log file again. "The phone is in idle state here. That's the normal state when the phone is not making a call, but it has already acquired a system."

Ravi leaned in and scanned the messages and was quickly confused. "What's a pilot? It mentions pilot and SEARCHER task a lot."

"Base stations have different pilot channels, which are like beacons. The mobile tunes to different frequency channels to find one with a strong RF signal. If it finds one, it tries to acquire the system by synchronizing its timing with the base station, decoding messages to learn what pilot channel it's on, and registering. It also learns how to look for other pilots that might be stronger. That's one reason for dropped calls when you drive. The base station is supposed to tell the phone about other pilots and base stations in the area so the phone can switch to a closer base station, but sometimes that doesn't work right and the call drops."

Ravi nodded, and began making notes from the log file. "There are a lot of sub-state changes and pilot changes going on in this file. I'm not sure what's normal."

"Neither am I. Let's try to reproduce the show floor again. Turn on logging in the phone so we can try to figure out what's happening."

While Ravi contacted Mike for specifics about the wireless coverage on the show floor, Oscar called Dave for more information about the configuration settings and got an earful about pilots, roaming lists, preferred networks, and the complexity and politics of roaming agreements among all the wireless service providers.

Of all the people at Omaha Telecom, he preferred working with Dave, who was a great resource and a no-nonsense embedded engineer. Low bullshit factor. Oscar picked his brain, supplementing the moderate knowledge of telecommunications standards he'd gained during a consulting gig years before. By the time Ravi had returned from configuring the base-station simulator, he had instructions on how to reprogram a roaming list in the phone.

Ravi led him to a wheeled cart holding the base station simulator they had on loan from Omaha. "Mike gave me the frequency and channel number and I configured the simulator, but I don't know how to tell my phone to acquire that pilot channel."

"Taken care of. I edited our roaming list to add the show floor signal - the strongest signal there is on channel 450, SID 4152." Oscar gave Ravi a shortened explanation of roaming lists, bands, channels, and SIDs as he connected a download cable to one of the two remaining phones.

As he prepared to launch the Austin application, Ravi suddenly interrupted the implicit countdown.

"Let me call Li Mei. She'll be mad if she misses the flames."

When she arrived, all three sat in a circle while Oscar resumed the test. Grasping the phone lightly in one hand, he was prepared to drop it on the bench if it started to overheat. Li Mei smiled in anticipation, and then reminded him to plug the charger into the phone.

Silence settled over the lab as they waited for the phone to power up. It acquired the new system and classified it as a preferred network; the PRL was good and the base station was correctly configured. Ravi relaxed microscopically. Shortly the phone displayed an exotic red sports car graphic and Ravi's welcome message.

So far so good.

As he waited, Oscar wondered if he imagined a slight increase in temperature, and refocused on the cell phone. Ravi and Li Mei both looked to him expectantly, but he had to shake his head.

The phone was working fine. It hadn't gotten warm at all.

"Pull out the log file," Oscar directed.

"This is pretty boring," Li Mei said. "There's not much searching or call processing stuff. Just a little pilot set maintenance. This is a lot different from the log Mike sent us." (See Figure 6-4.)

```
CALLPROC:    MS Idle State
BATT:        Battery update available msg
PILOT:       Pilot Set Maintenance
POWER:       Battery level medium
POWER:       Fast charging rate 75%
POWER:       Store charging state
UI:          Update icons/clock
UI:          Update icons/clock
RX:          Receive Overhead Msgs from BS
PILOT:       Pilot Set Maintenance
UI:          Update icons/clock
BATT:        Battery update available msg
POWER:       Battery level medium
UI:          Update icons/clock
UI:          Update icons/clock
BATT:        Battery update available msg
POWER:       Assess battery level
UI:          Update icons/clock
BATT:        Battery update available msg
POWER:       Battery level high
POWER:       Trickle Charge
POWER:       Store charging state
PILOT:       Pilot Set Maintenance
UI:          Update icons/clock
UI:          Update icons/clock

(Log file continues for several pages)
```

Figure 6-4 Debugging Log File from Idle Phone with Weak Signal.

Oscar compared the log files. She was right. The frequency of battery and power processing in each file was about the same, but Mike's log file showed a lot of searching and state changes. The battery was already low. Several times the phone looked for new channels and acquired new pilots before it found an acceptable system. He knew that each time it had to retune the synthesizers and find a new system, it drew additional current. All of those channel switches required that the phone stay awake longer to communicate with the base station, and turning on the transmitter to do so drained even more current.

"Ravi - change the simulator to some random channel we never use. Something not approved in the roaming list. I want to make the phone work a little harder."

A few moments later, he had a new log file and it looked a lot more like the idle log from the show floor. The searcher was working overtime and the phone communicated with the base station more. He let the phone continue to run as the screen cycled between "Searching . . ." and "System Acquired" while it struggled to find an acceptable system.

He closed his eyes, again trying to imagine that he felt the phone getting warm. Now, there was no doubt. Over several minutes he felt the case temperature very slowly creep beyond his body temperature. When he was sure the heating wasn't uncontrolled, he placed the phone in Li Mei's hand.

Her eyes opened wide in surprise and then a smile spread across her face.

They were halfway there.

After acquiring more liquid caffeine, Oscar had a frank conversation with Dave at Omaha Telecom about the possible source of the flames. Oscar felt he had enough ammunition to convince Omaha to step up to the plate and take a more active role in the debugging. Oscar's team had reproduced part of the problem; the over-warm phones needed more frequent recharging without even launching the Austin application.

Dave begrudgingly acknowledged that the phone was running at a pretty high utilization rate and was possibly starved for CPU cycles - maybe some code wasn't getting a chance to run - and he asked Oscar to rerun the test with the Austin app and capture a log. He wouldn't release access to the phone's source code, which Oscar was secretly relieved to hear. If the thermal problem happened again, they could at least take a look to see what the phone spit out the logging port.

"Before we try anything else, summarize what we learned from the last experiment and log file, Li Mei."

After a period of silence, she cautiously offered her assessment.

"Compared to the log with a good pilot, this one has too many things happening at one time. That could drain the battery and the charger has to run at 100%."

"But would that cause a fire?"

"I really don't know. Maybe if there was a short circuit. Or if the battery got too hot. But the POWER task is running and it checks the voltage and temperature values. I believe it uses this new information to change the charging rate if necessary. If it finds the temperature too high, it should prevent overheating by reducing the charging rate."

"True enough. In idle mode." Oscar nodded as he flipped his ID badge. "So what would happen when the phone gets even busier when Austin initiates a data call? The transmitter is on more then, too."

"Can we ask Mike to take a log file to capture the problem? Then we can see if the battery algorithm reports any problems."

"That is a possibility, but not compelling enough for me to ask Mike to singe his fingers again."

"I think we could make it happen here." Ravi stood up. "We need to make a data call, though. If we configure the base station to be a very weak signal, the phone might keep searching for a better one. Then we turn on data logging and carefully watch the temperature while the phone is being charged."

Apparently Dave's idea to reproduce the problem while logging was a contagious one, although images of exploding batteries with associated injuries still danced in Oscar's head. "Okay, but I control the experiment. I don't want anyone getting hurt."

Ravi scrambled to the base-station simulator and began punching buttons as Oscar opened the preferred roaming list file to change the system preferences. After he and Ravi had agreed on the good and bad system settings, they resumed their places to run the experiment. Li Mei had Austin and the pulse-ox sensors ready.

With a nod from Oscar, she initiated data collection and watched an LED on Austin blink happily; data was being received from the sensors. Oscar powered up the phone and waited several moments for it to acquire service. He noted with some satisfaction that it lost the system twice and had to reacquire it. That was a good sign that the system could be stressed out easily. At Li Mei's signal, he launched the mobile Austin app and waited for it to make connection with the monitor, and then initiated a data call to the base station.

Ravi peered over his shoulder at the mobile display, and told Li Mei when the base-station connection was completed.

Now they were in new territory. The phone was already warm and the charging indicator on the display flashed. Data continued to upload. Then Oscar noticed that the hourglass had stopped turning.

"Li Mei - is Austin still sending? We may be hung here."

"Yes, it is about halfway done with what I stored." She pressed a button. "It's still sending packets. I can see them counting up on the display."

Oscar stared at the phone, wondering if they'd lost the connection to the display board. The phone still seemed to be receiving data from Austin, but he wasn't sure if it was transmitting that data to the base station any longer. He pressed a random key and the phone beeped, but the display didn't change. Pressing more keys had no effect. As he leaned over to test the temperature of the charger, the phone suddenly became too hot to hold and he yanked the charger from the wall.

"That's it - turn everything off!"

Li Mei and Ravi both scrambled to shut down their respective pieces of equipment as Oscar popped out the battery using a stack of paper towels to insulate his fingers. Exhaling loudly and feeling his heart skip a beat at the prospect of a battery explosion, he kicked the battery across the floor and it skidded underneath a far bench.

"Well, that was interesting," Oscar said, calmly feigning nonchalance. "Shall we have a look at the log?"

Li Mei laughed. "You almost looked scared, Oscar! Did it get really hot?"

He admitted that it had. Luckily, he was spared crispy fingers. He motioned for Ravi to open the new log file, and everyone again crowded around the computer screen. (See Figure 6-5.)

The log started off about the same as the previous one, until the tasks supporting Austin began. The connection with the Austin Monitor was established and the Austin logo was displayed. They saw where Oscar had pressed the first key to initiate the data upload, followed by a reference to the hourglass display. The phone had then acquired a traffic channel to upload the data, and initialized the RLP and PPP engines to control the packet data transfer. Later in the file they saw where the APPS task initiated the data upload. Ravi had said that APPS supported applications like Austin or Phonebook. Throughout the file, they also saw references to pilot processing and to battery management.

Oscar motioned for Ravi to scroll down, but the file ended abruptly after the data upload had begun.

"That's it? What happened to the rest of it?"

Ravi vocalized the thought on everyone's mind. "There should be a lot of data transferred, and Oscar hit several keys at the end."

"I hit those keys after the display stopped updating, remember? Something happened in the phone that caused both logging and display updates to stop. But according to Li Mei, it apparently still allowed the data to be received. A partial hang."

```
PILOT:      Idle Handoff                          DISPLAY:    Launch Hourglass - Connecting
RX:         Receive Overhead Msgs                 CALLPROC:   Enter Access State
PILOT:      Pilot Set Maintenance                 CALLPROC:   Update Overhead Substate
BATT:       Battery update available msg          CALLPROC:   MS Origination Attempt Substate
UI:         Update icons/clock                    TX:         Origination Msg, SO33
RX:         Receive Overhead Msg                  RX:         Extended Channel Assignment Msg
CALLPROC:   Update Overhead Substate              CALLPROC:   Traffic Channel State
TX:         Registration                          CALLPROC:   Traffic Channel Init Substate
RX:         Receive Overhead Msg                  TX:         Send Preamble
CALLPROC:   Idle State                            RX:         BS Acknowledgement
POWER:      Battery level low                     UI:         Update icons/clock - charging
POWER:      Fast charging rate 100%               CALLPROC:   Forward Link Acquired
POWER:      Store charging state                  TX:         MS Acknowledgement
PILOT:      Pilot Set Maintenance                 CALLPROC:   Traffic Channel Substate
CALLPROC:   System Determination Substate         RX:         Service Connect Msg
SEARCHER:   Tune RF/IF                            TX:         Service Connect Completion Msg
CALLPROC:   Pilot Acquisition Substate            TX:         Service Connect Complete
SEARCHER:   Pilot acquired!                       BATT:       Battery update available msg
UI:         Update icons/clock                    CALLPROC:   Data Traffic Channel Ready
CALLPROC:   Sync Channel Acquisition Substate     RLP:        RLP Setup
RX:         Receive Sync channel msg              CALLPROC:   Pilot Set Management
CALLPROC:   Timing Change Substate                PILOT:      Soft Handoff
CALLPROC:   MS Idle State                         DATA:       PPP Setup
RX:         Receive Overhead Msgs                 POWER:      Battery level low
CALLPROC:   System Determination Substate         DISPLAY:    Hourglass - Connected
SEARCHER:   Tune RF/IF                            APPS:       Initiate data upload
CALLPROC:   Pilot Acquisition Substate            DATA:       Send Data Buffer
SEARCHER:   Pilot acquired!                       DATA:       Send Data Buffer
UI:         Update icons/clock                    CALLPROC:   Pilot Set Management
CALLPROC:   Sync Channel Acquisition Substate     PILOT:      Soft Handoff
RX:         Receive Sync channel msg              DATA:       Send Data Buffer
PILOT:      Pilot Set Maintenance                 DATA:       Resend Data Buffer
UI:         Update icons/clock                    DATA:       Resend Data Buffer
RX:         Receive Overhead Msg                  DATA:       Send Data Buffer
CALLPROC:   Update Overhead Substate              DATA:       Send Data Buffer
RX:         Receive Overhead Msg                  DATA:       Resend Data Buffer
CALLPROC:   Idle State                            BATT:       Battery update available msg
BATT:       Battery update available msg          DATA:       Resend Data Buffer
PILOT:      Pilot Set Maintenance                 CALLPROC:   Pilot Set Mana
PILOT:      Idle Handoff
POWER:      Battery level low                      (abrupt end of log file)
ACCESSORY:  Bluetooth device detected
DISPLAY:    New App Detected
APPS:       Austin app initiated
PILOT:      Pilot Set Maintenance
DISPLAY:    Austin Logo
ACCESSORY:  Austin connection complete
PILOT:      Pilot Set Maintenance
KEYPAD:     Process keypress                      Key:
APPS:       Austin upload initiated               RLP: Radio Link Protocol
UI:         Update icons/clock                    PPP: Point to Point Protocol
DISPLAY:    New App Logo                          SO33: Service Option 33 (Data call)
```

Figure 6-5 Debugging Log File from Phone with Weak Signal, on Charger, Running Austin.

Reader Instructions: The mobile phone is running a task-based real-time operating system. What could cause some parts of the phone to stop while others keep running? Based on these logging messages, what might have happened? How would you prove it?

Li Mei leaned back against the bench and squeezed her head as though it was ready to explode. "This is so complicated! If we look in the software, it probably has millions of code lines and we will easily get lost. But this log has not enough information."

Oscar looked between her and Ravi, noting that he shared her confusion, and he realized that neither of them had worked much, if at all, with real-time operating systems. He was still unsure what was causing the overheating, but he could use them as a sounding board as he brainstormed, as he had with Josie. At least they'd learn a little about RTOSes in the process.

"You both have worked on some of the microcontrollers that have a couple interrupts and a main routine with a big loop. And, Ravi, I know you know what happens when too much crap gets put in an interrupt service routine. Am I right?"

He noted Ravi's discomfiture and continued, "In more complicated systems like this phone, there are too many things that need to happen right away that are unrelated to one another. And some are more important, or more time-critical, than others. It becomes extremely difficult to balance all those needs with a few interrupts and a main loop.

"So, enter the real-time operating system, which is kind of like that monstrosity-of-an-operating-system running on your PC, except a lot smaller and more efficient." Ignoring their snickers, he continued without pause. "This RTOS allows many individual tasks to be defined, and the operating system controls which tasks get to run based on priority levels. For instance, we already know from the logs that the phone has a battery and a power task, a task to update information on the display, a task to control the data transfer, and a task that controls voice and data-call processing. Each has a unique priority level, and if two want to run at the same time, the higher-priority task gets to go first. In a preemptive system, the lower-priority task stops before it's finished to allow the higher-priority task to run. That allows time-critical things like sending data frames to happen before lower-priority things like writing some data to memory."

"Why isn't writing to memory important?" Ravi asked.

"It is, but it can wait a bit without ill effect. You have to consider how much a function will be hurt if it's delayed, and most things like memory or display updates can tolerate some delay."

"Is that what's happening here?"

Oscar pondered. "Not quite, I think. If the system got too busy, we should still see logging messages from the highest priority tasks. Instead, the log stops right in the middle of a command about pilot management. Strange."

He walked to his bench and began to type. "I am sending these log files to Dave. He offered to take a look if we could capture the phone misbehaving, and we have fortunately achieved a debugging suggestion on Li Mei's List - reproducing the problem."

With a final tap on the keyboard he added, "Now, we have to unravel what the evidence trail tells us. Since we don't get access to the code, Omaha has to join us in the fire."

Li Mei and Ravi sat in the cafeteria for a quick lunch break, exchanging what information each knew about the Austin system. Between them they had enough knowledge to just barely understand the problems the team faced, but neither felt experienced enough to generate a realistic solution.

"I am learning the wireless communications between the ECG sensor module and the Austin Monitor. They both use the same protocol but the data messaging is different," Li Mei explained. "But I don't think the phone application really cares what sensor the data comes from."

"It treats it all the same." Ravi mixed the food in his container and took a bite. "The sensor identification is in the data, but the raw data has headers and a checksum around it. That was one of the problems we had last night - something's wrong with some of the checksum calculations, and Josie thinks they changed it since the last release and now the phone might have a bug."

"Really? How did you get the phone software fixed?"

"We couldn't. The Omaha team doesn't have time to check it out, so we're doing the checksum wrong just to get through the show."

He leaned in to ask confidentially, "Can you believe they are messing up something as important as the checksum?"

Li Mei leaned closer. "I can't believe you guys put the wrong message length in the header! That's worse!"

He leaned back, embarrassed. "Yeah, we did mess that up in the middle of the night. But I found it pretty quickly. Before Josie did."

She decided not to bait him. Instead, she changed the subject to ask about his signal-processing algorithms for the medical data, which he was much happier to explain. After they finished lunch, Li Mei swung by her cube to surf the web for information about overloaded RTOSes and sources of performance problems in mobile phones.

Although Oscar knew little about cell phones, he did know that acquiring and keeping good systems was a high priority; otherwise the phone dropped calls. That got him thinking - how many tasks were running at the same time? 20? 40? Many more than the handful they saw in the logs. All vying for precious system resources.

Some might not be getting a chance to run. Could that affect the phone's ability to acquire a good system and keep it?

He supposed that if the phone were in a car moving from one base station to another, it might not find the new base station if it were too busy doing other things.

But they weren't moving the phone more than a few feet in the lab *or* at the show.

In the lunchtime quiet of the lab, he gathered his thoughts to organize everything he knew so far. When the phone wasn't too busy with system acquisitions, Austin worked fine and the phone didn't get warm in or out of the charger. When the phone had to continuously reacquire new systems, it got very warm just sitting in the charger *or* out of the charger. But when the phone was searching for a system while in the charger and Austin was launched, flames ensued.

That meant the only untried combination was searching for a system, out of the charger, launching Austin.

He patted down the front of his shirt and tucked it neatly back into his jeans where it had escaped, wondering if performing that final test would gain him any additional information. It could exonerate the charger as a contributing factor.

When he couldn't keep the permutations straight in his head any longer, he gave up and made a truth table of what they had tried and learned. (See Figure 6-6.)

Inputs				Symptoms	
Constant System Reacquisitions	Charger In Use	Austin App Active		Phone Very Warm	Flames
No	No	No		No	No
No	Yes	No		No	No
No	No	Yes		No	No
No	Yes	Yes	⇨	No	No
Yes	No	No		Yes	No
Yes	Yes	No		Yes	No
Yes	No	Yes		???	???
Yes	Yes	Yes		Yes	Yes

Figure 6-6 Truth Table Comparing Inputs with Observed Symptoms.

Drawing up the table confirmed his suspicions about the acquisition problem; it was independent of the charger. His mind kept returning to the final lines in the log file. The BATT task had information to report, but the corresponding entry from the POWER task was missing.

His gut made the decision for him - they would run the extra test.

But his focus had changed; he was less interested in whether the phone would smoke and more interested in the log file. Would it terminate early as before or would it give them a full picture of the internal processing? Would all the tasks run or just some of them?

Because he had an idea. A nasty one.

> **Reader Instructions:** What does Oscar suspect? What do you think the results of the test will show?

When Oscar returned to the lab, he had a resolute expression on his face. Li Mei was sure he had figured something out and wasn't happy about it. He pulled up a stool to face them both and explained the conditions for what he hoped would be the final test.

They would repeat the flaming Austin experiment with the phone out of the charger. He told them he believed it would not singe fingers, even with the very weak base-station signal.

Li Mei wondered why he thought so, but already felt swamped trying to understand tasks and overloaded processors and mobile phones searching for systems. She decided to quietly observe. Ravi fidgeted and tried to ask why, but Oscar did not elaborate.

"We will try very carefully to repeat everything we did before, but with no charger," he instructed. "Down to the individual buttons I pressed after the hourglass stopped spinning. Got it?"

With nods all around, they initiated the test with quiet concentration, each handling their specific task. Li Mei couldn't see the display on the mobile phone, so she watched Oscar's face for clues.

After a minute Ravi lifted a finger; they must have launched the Austin application.

The Austin Monitor in front of her began to blink and she waited patiently as it sent the data.

Oscar looked expectantly at the phone's screen. "Hourglass still spinning."

The vigil continued. After a time, Oscar pressed several buttons on the phone, continuing to peer at the screen. Finally, just as she saw the Austin Monitor stop blinking, he nodded slightly. "Okay, that's it. Data uploaded successfully."

"It worked? The Austin Monitor says upload complete as well." Li Mei finally spoke. "It's not too hot?"

"Warm but not too hot. This bug requires the charger to be connected. Now, we know that for sure. But a second piece of evidence is that the phone has to be working hard doing other things, like Austin data transfer or searching for service." He set the phone on the bench and turned to face her.

"The phone is overloaded, and I wager we will find a complete log file. We will still find several reacquisitions of the system, and data resends, but no references to the battery or power tasks. Check it."

Ravi was already processing the log file and shortly had it displayed on the computer monitor. Li Mei joined him at the monitor and watched as he scrolled past the call setup and through several pages of data transmission and retransmission.

Oscar was right. This file looked complete up through the last key press.

She marveled at how he could predict the results of tests, as if he had already run them himself. She wondered if she would ever feel confident enough to make such predictions in front of others.

"How did you know?" she asked him.

"The last few logging messages from the bad trial got me thinking." His face took on the faraway look she was learning to associate with mental debugging and concentration; it was a look that conveyed his fascination with the technical ways machines could fail.

"That last entry refers to storing information about pilots. That's the last message, and that message is truncated. Something went wrong *right there*. At that point, the phone is trying to send a lot of data over the air. Those data-send operations should be pretty high priority, and the phone is doing a lot of searching and transmitting. That's draining the battery."

"The battery was already warm," Ravi offered. "Maybe that's when it detected the battery was getting too warm and decided to reduce the charging rate below 100%."

Oscar raised his finger sharply. "Yes. I believe so. Very good, Ravi." He remembered to emphasize his praise by pausing to look Ravi in the face. "It sensed a thermal event in progress and attempted to avert it. However, it never actually reduced the charging rate, which ultimately caused the phone battery to overheat and led to the fire."

"But why wouldn't it reduce the charging rate?" Li Mei asked.

"A very cool but nasty condition called priority inversion." Oscar looked deadly confident in his assessment. "Ever heard of it?"

Both shook their heads. Li Mei opened her notebook and wrote down the phrase.

"Priority inversion occurs when a task with a high priority is unable to run because a task of lower priority is blocking it." As he spoke, Oscar's expression hardened, as if he had accepted the challenge that had materialized before him. "Normally that can't happen, but with three different tasks and a shared resource like physical memory, it's possible and actually pretty common."

He rose and walked to the whiteboard. "Here's what can happen. We have three tasks, each with a different priority level." He drew a horizontal line for each task (see Figure 6-7). "Say the lowest-priority task runs first and locks a resource like memory because it needs to store some variable." He added an arrow from the low-priority task to the resource below it. "Physical resources like memory should only be accessed by one task at a time, so if the resource is available, the calling tasks will lock it so no other task can access it. In this case, the low-priority task is able to lock the resource.

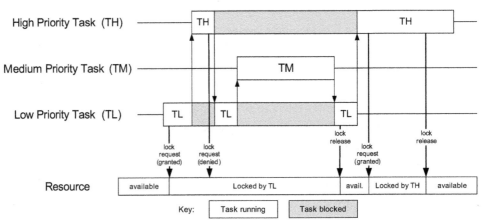

Figure 6-7 Priority Inversion.

"Next, say a high-priority task comes along and stops the low-priority task because it has a higher priority level. It runs for a while, but then wants the resource, too. It makes a request, but since the low-priority task already has it locked, the request is denied. The high-priority task must wait until the low-priority task is done with it."

Li Mei interrupted him. "I don't understand. If the high-priority task has a higher priority, shouldn't it get to run?"

"Normally, yes. But in the case of a physical resource, it can't interrupt a hardware action in progress." He grappled for real-life examples. "Like reading a port before updating the state of the pins, or confirming a memory-write operation. And since the high-priority task can't do anything until it can access the resource, it has to wait. That's usually no big deal because resource accesses are designed to be quick and efficient. It's like disabling interrupts for just a moment to protect a critical segment of code." Oscar paused. Li Mei's grasp of the situation was tenuous, but she nodded for him to continue.

"So the low-priority task goes back to running, but before it can finish with the hardware resource, say a third task of medium priority comes along. Keep in mind the high-priority task is still waiting. The medium priority task stops the low-priority task from running because it has a higher priority, just like the high-priority task did before. And *unlike* the high-priority task, this medium-priority task *doesn't* need the resource, so it's allowed to run to completion. This medium-priority task prevents the low-priority task from releasing the resource.

"That's why this is called priority inversion - the medium-priority task gets to run and the high-priority task doesn't."

Oscar turned to look at the board himself and uttered quietly, "This is it. This is just amazing."

Li Mei copied his drawing in her notebook while she tried to think about how this could apply to their situation. She recalled his earlier comment and asked, "Could physical memory be the resource that gets locked for us? Maybe the last line of the log was trying to store something about pilots to memory."

"I suspect so." He turned back to face her. "But the log doesn't show every task and we don't know the relative priorities of these tasks. Let's brainstorm based on what we *do* know. Anyone care to guess which is the high-priority task in this scenario?"

She looked back to the previous log file. "Not all the tasks are logged?"

"They can't turn on all the messages or the phone wouldn't run." Oscar had confirmed with Dave that the results were filtered. "They just enable what is relevant at any given time."

Ravi pointed to the log file. "I assume that either the POWER task or the logging task locked the memory resource. DATA sending is probably high priority, and maybe the CALLPROC task sneaks in to handle the weak pilots and stops everything from running." Before he could continue with his idea, Oscar interrupted him.

"Ravi, why would a task sending data over the air need to write anything to memory? That's a critical task that requires adherence to previously negotiated timing with the base station. It's reading from a buffer, not writing to FLASH." Oscar looked annoyed. "Think."

He faced Li Mei. "We need two tasks trying to write to memory at the same time. Which tasks could they be?"

"I think the logger and the POWER tasks," she offered.

"Correct. Now which gets the resource first? Be careful," he cautioned.

She looked back at the log file, which showed that CALLPROC was the last task to run. But she also knew the logger task wasn't explicitly listed. After running each scenario in her head, she made a decision.

"I think the logging task gets the resource first and is storing a message about the pilots. Something about all these system acquisitions."

"Yes. The logger is the low-priority task in this inversion scenario. Now to the high-priority task. Who also wants access to memory? Think about what usually happens after the BATT task reports information from the battery."

Ravi scanned back through the log file. "The POWER task always responds a bit later. It runs the algorithm to decide if the charging parameters should be changed. Here at the beginning," he pointed, "it reads the battery voltage as low and turns the charger on 100%. At other places in the file, the battery is still reported as low, but it doesn't change the charging rate because it should still be at 100%."

"Good. Now what about the end of the file?" Oscar coached him. "What does the POWER task do in response to the last BATT message?"

"It doesn't do anything."

"Again, be careful. The log file doesn't *show* anything because it stopped in the middle of logging information about pilots." He tried to explain logically. "If POWER is the high-priority task and it needs to save a new *lower* charging rate but is *blocked* behind the logging task, what happens?"

It was all starting to make sense to Li Mei and she jumped in with excitement. "If it updates the new charging rate *after* saving the new rate to memory, then the charging rate is still stuck at 100% because the memory write never happened because anything with a middle priority interrupted the logging task. Like all the pilot searching and data sending."

Ravi added, "And then the battery continues to overheat while the phone is too busy doing other stuff. I understand now."

Oscar turned back to the whiteboard. "So we propose that the high-, medium-, and low-priority tasks in this scenario are POWER, DATA, and LOGGER." He added the task names to his diagram before pulling up a stool. "Definitely could've caused the flames, and I'm calling Dave right now. If they're running the POWER task at a high priority to prevent battery problems, they shouldn't be saving anything to memory - just checking the voltage and temperature and adjusting the charging rate. I'm amazed at this architectural wonder."

"So what do we do now?" Ravi asked.

"You and Li Mei unplug the charger and rerun the complete Austin test about 50 times with some large data files and the weak pilot and take logs. Do it with weak and freshly charged batteries - all permutations except the flaming configuration. I want some confidence before I tell Mike that he can go ahead and start demoing Austin again, sans charger."

Li Mei sat at her desk, entering into a spreadsheet the results of the 56 Austin data trials that she and Ravi had performed, when her email notification beeped. It was from Josie, reporting a smooth flight and a full lunch spread, with homemade fried chicken and buttercrumb apple pie, waiting as soon as she reached her grandmother's apartment. Josie asked if the phones had arrived at the show okay and Li Mei suddenly realized that Josie had no idea about the fire!

Tapping out a quick email, she outlined the events of the day and added Oscar's insight about priority inversion.

"Don't worry, your code is good. Our tests were successful, so now we can demo the Austin application and Monitor at the show and Mike will be happy," she typed. "Just no logging and no battery chargers until they solve their priority-inversion problem. I also learned about real-time operating systems and task priorities, and how to debug a problem by looking at log files. It is a different kind of detective work. It's especially good that we didn't have to look at one million lines of code."

The next morning Randy and Tom joined the team in the embedded lab, and Randy read the email update from his management. Based on input from both Hudson Technologies and B&L Omaha Telecom, the Austin Medical Monitor was currently a popular demonstration at the World Wide Telecom Show with no adverse events. Everyone at the booth had been cautioned against using the charger during demonstrations, and they'd charged a stack of spare batteries overnight to keep the demos running smoothly.

Oscar was relieved to note that Randy's report contained much of the same information and conclusions as he had gotten from Dave and Mike. Dave had confirmed that unbounded priority inversion was a definite probability and that Oscar's team had pegged the relative priority levels for the three tasks in question. A quick run with a performance profiler under the show floor conditions had revealed that the phone was overloaded and several lower-priority tasks were not able to run.

"Another thing Dave told me is their system-searching algorithm has some problems. If it can't find a good system, it keeps searching repeatedly. It's supposed to take a break to keep from draining the battery; that's one of the reasons the phone gets warm in bad pilot areas."

"But the show floor had a good strong signal. Why couldn't they use that one?" Randy asked.

"Oh, this one's classic," Oscar grunted. "Operator error. They put the show floor system on the roaming list, but no one bothered to make sure it was classified as a preferred system. Seems it was listed as a negative system that the phone wasn't allowed to use for service."

Tom groaned. "So that's why it was searching like mad. Dave never got back to me about the antenna connections, but now it all makes sense.

"This reminds me of that telecom trade show in Singapore several years back. We had a new phone out, back when one of the digital cellular standards was first commercialized."

Oscar found himself nodding at the forgotten memory. As Tom continued his story, Oscar realized how little perspective Ravi and Li Mei had, being so new to the field.

"Since the service wasn't live in Singapore yet, we had these large tractor trailers carrying base stations lined up around the hotel where the show was held. They ended up drilling a hole through the outside wall of the hotel to snake the antenna cable into the building to the show floor." Smiling, Tom shook his head at the thought. "That must have cost a pretty penny."

"Did the phones work?" Li Mei asked.

"Pretty well. We got good publicity from the show as a new player in the field." Tom added.

"How about this one," Oscar offered. "We had a trade show overseas, back before we could easily download software over the internet directly into the product. The software still had major bugs and the home team was working 24/7 while I flew to the show." He leaned back against the bench and hooked his feet behind the stool. "To program one of the processors, we had to use full-sized card in a custom computer. No big deal in the lab, but we had to ship the entire system with the monitor to the show so I could load the fixes. I spent several hours download-ing new software at 2400 baud over the phone lines in the middle of the night to get the lowest long-distance rate. Every new version of code cost a couple hundred bucks to download."

"You couldn't just take a laptop?" Ravi was surprised.

"Nope. No PCMCIA cards or USB devices for that processor back then. Times have certainly changed. Now I can log onto the server and download software for

Austin through the USB port. Very cool. And to top it off, the monitor didn't survive the trip home, and the computer was stuck in customs in Alaska for almost a month." Oscar added. "But at least the trade show was a success."

Li Mei shifted in her seat. She looked anxious to ask a question but was hesitant to interrupt the managers swapping war stories. Oscar let her off the hook with a nod.

"You said before that priority inversion is pretty common. If that were true, how come things don't catch fire or cause huge problems more often?"

Oscar shifted gears and Tom got up to leave with a wave. He noticed that Randy had already gone. "Good question. Most of the time priority inversion resolves itself without anyone knowing. Remember the picture I drew on the whiteboard with the three tasks?"

She nodded.

"Let's say that the medium task finishes running. Then the low-priority task resumes because the high-priority task is still blocked waiting for the resource. As soon as the low-priority task releases the resource - for example, it finishes writing something to memory - the high-priority task can then lock the resource and run. Inversion resolved and no one's the wiser."

"So how will Omaha Telecom stop that priority inversion from happening?"

Oscar raked a hand back over his head and stared at the ceiling. "They can attack this problem in different ways. First, they know the system is overloaded, so they need to cut out debug logging and fix the searcher so it isn't off looking for systems constantly. You'd be surprised how many products ship with debugging code still running because someone forgot to turn it off."

"Second, I'd challenge Dave on why the battery algorithm has to save its charging rate to memory in a high-priority task. If their operating system has messaging, they could just send a message with that information, to be written when the system has time. Or save it in a variable."

"But wouldn't that prevent the algorithm from changing the rate?"

"If it does, then it's a crappy design. Safety-critical actions need to be performed at a high priority without being blocked."

Li Mei offered, "I read on the internet about something called priority inheritance to solve this problem. Can the phone do that?"

"Depends on the RTOS. If it can, then the priority of the logging task would be elevated to the priority of the POWER task so it could finish logging and release the memory resource. Then POWER would save its new charging-rate information and then be able to reduce the charging rate before the fire occurred.

"Ultimately, they should have designed the system to avoid priority inversion in the first place. Allow only one task to access the resource, or make all tasks that need the resource share the same priority level. Another option is to use message passing to the resource task rather than resource locks."

He raised an eyebrow and concluded, "But whatever they've got now, they'll probably just muck with it a bit to make it work. Could be too late to make major changes."

"So all of this ended up to be Omaha's problem, and we got forced into debugging it for them," Ravi said.

"Be careful about being so black-and-white about things," Oscar told him. "Sometimes debugging someone else's problem ensures your product gets to market. Or, in our case, it helped our product get great customer exposure at an international trade show. In saving their skin, we saved our own."

Additions to Li Mei's List of Debugging Secrets

Specific Symptoms and Bugs

- RTOSes using preemptive task scheduling can allow priority inversion if tasks of different priority levels can access the same common system resources, such as hardware.
- Remember to turn off debugging code before shipment!

General Guidelines

- Identify the set of inputs that causes unacceptable behavior, then find what must be changed to make the behavior acceptable.
- Periodically stop and summarize findings.
- Leverage collaborators including vendors and customers.
- Consult a guru who has seen many different types of bugs before.

Chapter Summary: The Case of the Flaming Phones (Difficulty Level: Moderate)

A dangerous trade show demonstration brings the team together as they struggle to reproduce the problem offsite. When the Austin Home Medical Monitor and the B&L Omaha Telecom phone connect to transmit medical data over the air, the phone catches fire. Using reports from the show and observations of different mobile phone behaviors from logged messages, they identify the combination of conditions that leads to the fire. Priority inversion from an overloaded processor prevents the phone from turning off a battery charger, allowing a thermal event to occur unchecked.

The Problem Symptom(s):
- The phone caught fire while running the Austin application and connected to the battery charger.
- All phones were warm and needed to be charged very frequently.

Targeted Search:
- Bad batteries were suspected, but new batteries from the factory did not prevent the problem.
- Unable to access the proprietary software, the team reviewed debug logging files produced by the phone.
- A truth table of inputs (charger, Austin, etc.) isolated the exact conditions that invoked the problem.

The Smoking Gun:
A combination of charger, weak signal, logging enabled, and Austin caused the thermal event. The log file was truncated in a manner consistent with priority inversion.

The Bugs:
A combination of events led to the fire.
- The strongest wireless signal on the show floor was listed as a negative system in the phone's roaming list.
- Phone searcher searched for a better system too often, which drained the battery and system resources.
- Priority inversion occurred when the high-priority POWER task attempted to reduce the charging rate but was blocked from saving the new charging rate by the lower-priority logging task. Medium-priority data transfers prevented the logging task from releasing the resource.

The Debugging Method Used:
- Interviewing the problem reporter.
- Reproducing the problem.

- Brainstorming with other groups (hardware, marketing, B&L Omaha Telecom).
- Creating a truth table to categorize inputs and symptoms.

The Fix:

Three changes were made to avoid priority inversion at the show.
- Updated the roaming list to classify the strong system as preferred.
- Disabled logging.
- Removed chargers from the show floor and used fully charged batteries.

Verifying the Fix:

- A series of tests showed that removing the charger allowed data transmission without fire.
- B&L Omaha Telecom verified less overload when logging was disabled and the roaming list was properly configured.

Lessons Learned:

- Adding features to an existing product requires full testing over the wide range of conditions under which the product operates.
- Devices starved for system resources don't always show obvious symptoms or degrade performance nicely; instead, unexpected catastrophic failures can occur.

Code Review:

Even though the software was proprietary and unavailable, the team could predict priority inversion performance problems that prevented proper operation. Had they been granted access to the software, they might not have solved this mystery in time.

What Caused the Real-World Bug? When Bethlehem Steel's measurement instrument approached the end of the track, a limit switch, sensed by a polled I/O, was designed to stop the unit from traveling off the end. During testing, it was found that the Z80 microprocessor was so burdened that the polling loop ran too slowly. To solve the problem and prevent a safety issue, an interrupt service routine (ISR) was added to handle the switch. The polling code was commented out.

But no one had tested the ISR properly. On the very first run, the instrument ran off the track, stopping only when the 440-volt motor drive wires failed, arcing and causing a small fire. [2]

References

[1] Metzger R.C. (2004), *Debugging by Thinking: A Multidisciplinary Approach*, Boston: Digital Press, p. 9.

[2] Ganssle, J., Personal communications, August 1, 2006.

Additional Reading

Renwick, K. and Renwick, B. (2004), "How to use priority inheritance," *Embedded Systems Design*, accessed from *http://www.embedded.com/showArticle.jhtml?articleID=20600062*.

3GPP2 TSG-C, (2005), "Upper Layer (Layer 3) Signaling Standard for cdma2000 Spread Spectrum Systems," 3GPP2 #C.S0005-D V2.0.

Chapter **7**

The Case of the Rapid Heartbeat: Meeting the Spirit of the Requirement

J osie hunched distractedly at her computer, half reading her morning email and half letting her brain unravel the lingering data-transmission problems with the Austin Monitor.

A new embedded poll caught her attention.

How much time do you spend debugging your embedded system?
  ○ It works fine first time it's powered up,
  ○ Less than 25% of the development time,
  ⊙ Less than 50% of the development time,
  ○ Most of the development time, or
  ○ Why test? We just ship it. [1]

She smirked. Sometimes the folks in marketing and management have *no idea* how complicated these systems can be. Just because the user interface works doesn't mean the entire *system* is stable.

Well, she amended, everyone certainly got a taste of barely stable software at the World Wide Telecom trade show last month. That bug was pretty spectacular but completely preventable if they'd been able to complete basic integration and some field testing.

183

It reminded her of another near-disaster she inadvertently caused by "cleaning up" the software right before it was supposed to ship. Shortly after her code beautification, she'd received a frantic call from manufacturing that the measurement devices were suddenly reporting wildly wrong answers. She'd simply wanted to take the difference between two long integers and then cast the result to a char and store it. Even though she had properly placed the parens around the subtraction operation to tell the compiler to perform that operation first, the compiler, in its infinite wisdom, had decided to cast the first long to a char and *then* proceed with the subtraction. That little doozy had kept her up for hours until she disassembled the C code to find that the problem was in the compiler, not in her software.

She eventually found eight other bugs in that compiler. It was a hard lesson about fallibility of the tool chain, and she'd vowed to never forget that the tools could contain bugs as well.

Shipping prettier code had been a nice idea, but she'd never do *that* again at the last minute.

As she pondered the poll, she heard a gentle tap on her cube door and looked up to see Li Mei peeking around the corner. She looked agitated, which was unusual.

"Hi, Li Mei, come in. What's up?"

"Hi, Josie." Li Mei entered and slipped into the guest chair. "I have a problem with the electrocardiogram part of the Austin Monitor. I am using your test plan and testing my side by taking my own ECG, but after the data gets to your side, it says my resting heart rate is 180 beats per minute.

"That's not right," she appealed, tugging at the ECG leads that extended from under her blouse. "Do you have some time to help me?"

"You have something working already?" Josie was surprised that Li Mei had made progress so quickly, and decided she was game for a distraction. "I was just working out the bug where the Austin app keeps requesting a high data-rate channel for our very low-rate data, but I'm stuck. Let's go look at your ECG."

Josie walked with Li Mei back to the lab. Even though it was still early, the lab was loud with the hum of test equipment and classic rock. Tom had a boom box precariously balanced atop unused function generators, and a stack of CDs to compensate for the nonexistent radio reception in the center of the building. Music seemed to be a universal necessity for hands-on work, although the selection of music sometimes generated a heated but friendly rivalry between the groups.

"We tested all the hardware stuff and it's working." Li Mei reconnected the ECG leads to the breadboard on her bench and turned on the power supply. Almost immediately, a green LED on the breadboard started to blink.

"Now you can see my heart rate here." She pointed to the LED, then touched two fingers to the carotid artery in her neck. Josie waited as Li Mei counted softly to herself.

"Yes, that is blinking one time for every one of my heartbeats. But the Austin display says 180 beats per minute. That is too fast. My LED is correct."

Josie smiled to herself, refusing to be baited, and stepped back into the mentor's role.

"Okay, let's stop and classify the symptoms. It *does* look like the LED is beating slower than 180 BPM. What is your actual heart rate?"

"This LED is not 180 BPM, Josie." Her voice rose. "I am not 180 BPM!"

"Li Mei, one step at a time, okay? Let's gather real facts and not make any assumptions." Josie stood back to look at her more closely. "Why are you upset? Heck, for all I know your heart rate *and* your blood pressure are off the charts right now."

"Okay, I am mad," Li Mei admitted. She dropped onto the stool and took a deep breath. "My final design isn't due for a while, but I wanted to make this work the first time. I want to be a very smart team person . . ." her voice trailed off. "The wireless communications part between my sensors and the Austin Monitor is hard. I think I didn't do it the best way."

"That's okay." Josie sat down slowly. Li Mei was stressed out but Josie didn't think it would last. Josie felt some kinship with Li Mei; she was also very hard on herself. She hoped the challenge to solve the problem would win out.

"Let's characterize it with a stopwatch for now, just to get an idea how big the discrepancy is. How about I count the LED flashes and you count your heartbeats. You say 'Go.'"

On Li Mei's mark, both began counting silently. At the end of one minute, Li Mei reported 91 heartbeats.

"I got 90. Same thing. So the LED seems to be the same as your heart rate, and definitely not twice that rate."

> **Reader Instructions:** Before continuing, take a look at the block diagram in Figure 7-1, and recall the Austin system elements in Chapter 6, Figure 6-1. Does this test give you useful information about what might be wrong?

Josie continued, "Now show me where the LED is in the circuit. It's driven by the microprocessor, right?"

"No, the LED is in the signal conditioning, before the microprocessor."

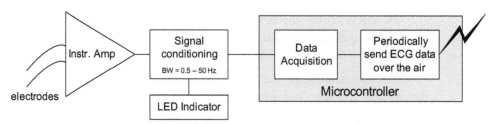

Figure 7-1 Block Diagram of ECG System.

Josie smacked her forehead. "Well, then the LED *should* be the same as your heart rate unless the signal-conditioning circuit is the problem. I thought the LED was driven by the micro, so that was probably a dumb test we did. Bad assumption on my part. Where's your block diagram?"

Li Mei dug through a small stack of folders and extracted one carefully labeled ECG DOCUMENTATION. She offered the block diagram (Figure 7-1) and a schematic to Josie.

Josie shrugged; the circuit was straightforward. The ECG signals first passed to the signal-conditioning block. The first stage was a low-power instrumentation amplifier with high common-mode rejection, followed by a low-pass filter to remove high-frequency noise and 60-Hz line noise. Then the signal was high-pass filtered to restrict the frequency spectrum from approximately 0.5 Hz to 50 Hz. The amplified and filtered ECG signal passed to the second block for analog-to-digital conversion. The analog-to-digital converter (ADC) was physically a part of the microprocessor, although Li Mei had drawn it as a separate element. That was fine for clarity's sake. The filtered ECG signal also fed into a separate LED indicator block that contained a simple analog comparator. When the ECG signal exceeded a preset threshold, the output of the comparator went high and the LED turned on.

"Well, at least the hardware appears to be working correctly. Sometimes that's half the battle. Now explain to me what the micro does with the A/D data."

"Inside my code, I have the ADC programmed to take a sample every 1 millisecond by using the timer. When the TIMER runs, I acquire one data point and store it in a temporary buffer and set a flag that new data is ready. In the main loop of the program, I wait for the new data flag and then send the data to you. It is not complicated but for some reason it's not working, and because of this I am very frustrated! Can you see if my data is getting to you?"

"In a minute. First, I want to understand this a little more without looking inside the box." Josie checked some of the dials on the Austin monitor. "As a detective, I have some questions: does Austin always display 180 BPM, or does the number change? Just that one piece of information will point us to very different parts of the software. Also, if your heart rate goes up, does Austin's?"

"That's true . . ."

"Remember, Austin already works for the pulse ox and blood pressure sensors, so it's more logical to look in your software before looking in mine. Don't just assume it's someone else's code."

"You are right. I'm sorry, Josie. I didn't mean to imply that your code has bugs."

"That's okay." Josie waved it off. "I'm sure Austin still has a few. But I notice that, as we sit here and talk, Austin's heart-rate display is down to 160 BPM. Has yours decreased as well?"

"Well, maybe." With her head still bent, Li Mei peered up abashedly through her lashes and admitted that she had not been using her carefully prepared list.

By lunchtime, they had determined that the Austin display tracked Li Mei's heart rate and the LED by roughly a factor of two. Even if Josie unplugged from Austin all the sensors she was using - pulse ox, blood pressure, and her new foot pressure sensor - Li Mei's displayed heart rate had the same error. Josie was relieved to find that her real-time operating system wasn't overloaded with multitasking among the sensor drivers, as the trade-show phone had been.

Josie walked around to her bench and asked Li Mei to snake the debugging cable between their benches. "Another black-box test we can run on your software is to check exactly what data you are sending me. That's easy enough to do.

"Okay, all Austin does right now is wait to receive a message, parse the data, and then store it depending on what type of data it is. Your messages should start with the ECG_DEVICE Message ID byte so I know how to interpret it." Josie paused to look at the screen, and then executed several quick keystrokes.

> **Reader Instructions:** What do you think will be displayed by Josie's code in the Austin Monitor? Look at Li Mei's flowchart algorithm in Figure 7-2. Describe the data she is sending (content and timing).

"Okay, I'm running in the debugger. Go ahead."

Neither spoke for several seconds as the LED began blinking again.

"Well, my program computes the heart rate fluctuating between 156 and 162 BPM." Josie allowed several seconds of data to be stored and then halted her program. Shortly, she had the raw data imported and plotted in a spreadsheet with a few annotations (shown in Figure 7-3 [2]).

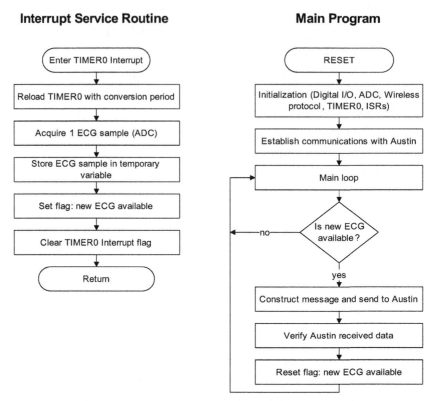

Figure 7-2 Software Flowchart for the Wireless ECG Sensors.

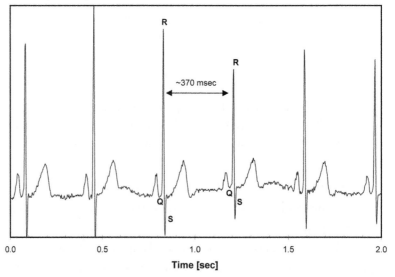

Figure 7-3 Li Mei's ECG Signal Received by Austin.

She twisted the monitor around for Li Mei to see. "This looks like a normal ECG waveform. If I take a rough time measurement between two QRS peaks, it's about 370 milliseconds, and that's . . ." She scribbled an equation on her pad and then punched the numbers into her calculator.

"162 beats per minute." She showed Li Mei her math. "See? Austin knows you're sending ECG data and it knows the time interval between each data point is 1 millisecond. It does about the same calculation I did, but with a little more intelligence to automatically detect the QRS complex during noisy data and motion artifacts."

$$\frac{1\,beat}{.370\ \cancel{sec}} \times \frac{60\ \cancel{sec}}{1\,min} = 162.2\ \frac{beats}{min}$$

"So, Austin computes the heart rate correctly based on these data." She switched gears. "Try something for me; jump up and down."

In response, the Austin display rose and leveled off at nearly 200 BPM. "The data are coming in pretty consistently."

"Hey, no workouts in the lab!"

Josie looked over her shoulder to see Tom walk into the lab.

"Shut up, resistor-boy, we're doing real work over here. Not sitting around drawing pretty geometries on the computer."

"Ouch, I'm wounded! I didn't realize that developing software was like a Habitrail - if you jump up and down faster does the code run better?"

Josie shot him a look and contemplated throwing foam sensor packs at him.

"Ha ha. Funny. If you guys had spec'ed out a faster processor we wouldn't have to use the exercise wheel, now would we?"

Tom dodged the blow as he rounded the corner back into hardware. "Can't blame this one on hardware. Seeyaaaaaaa!" His voice trailed off in a falling pitch until they heard silence, followed by a loud smack against the wall. His signature piece.

"That guy's a piece of work."

Li Mei was already laughing. "What's a Habitrail?"

Josie chuckled and propped her foot against the edge of the table. "Oh, Habitrails! Do you know gerbils and hamsters? Popular pets for kids?"

"Oh, I know now. Mazes made out of different color plastic tubes, right?"

"Yup. Tom was being a smart ass about gerbils running like mad on the exercise wheel and not getting anywhere. Makes me think we ought to take care of Tom next." Josie stood. "Speaking of which, I've gotta finish the plans for tomorrow morning. Why don't you make a list of everything we've learned, and then, well, develop some hypotheses about your bug, okay?"

A conspiratorial smile crept over Li Mei's face as understanding set in, and Josie noted that she was in a much better mood.

"Yes, I'll make a list. And Ravi and I will be ready for tomorrow morning."

The browser was still pointed to the unanswered embedded-systems poll. Dragging her thoughts back after a long swallow of hot coffee, she tried to estimate the relative development and debugging times in her head.

Debugging some projects was harder than others, especially when the team inherited legacy code like Ravi's animal interface. Or with more complicated multitasking and multiprocessor systems, like the entire Austin Medical product line. But when they were able to control the design process from beginning to end, as with the Austin Monitor itself, the debugging phases ended up much shorter.

Making up her mind at last, she clicked on "Less than 25%"and then shook her head in amazement at the results page: 68% of respondents reported more than 25% of the time. She was in the minority.

Wow, she thought, maybe some of the techniques that Oscar is pounding into our heads are actually paying off.

With that realization, she took another gulp of hot coffee to hide a twinge of misgiving at what she had cooked up for him this morning, but a small giggle floating over her cube wall quickly dispelled her hesitation. She stepped up on her guest chair to peer over the wall into Li Mei's cube.

"Shush - he'll hear you! Usually no one's here this early; if he hears anything, he's gonna know something's up. So be quiet!"

"I'm sorry, Josie, it's Ravi, not me!" Li Mei wrapped her arms about her and perched on the edge of her chair, annoyed with Ravi and worried that Oscar might be mad at her.

Josie was surprised to see that Ravi's grin carried an underlying tinge of vengeance, and she wondered if the contention between the two men was worse than she thought.

"Okay," Josie pursed her lips and glared at Ravi. "Just sit there without making a sound, if that is *at all* possible."

7:05 a.m. It would be any moment now.

With a final conspiratorial look around, she dropped back into her chair to wait, knowing that she had come up with the best practical joke yet.

Everyone hated working in the noisy cubicle farm, and Oscar regularly complained that to concentrate and get any work done, he needed a door to shut. By coincidence, she found out from Kathy that maintenance had extra cube partitions.

With a spare partition and two partition connectors, the doorway access to Oscar's cube had been eliminated. The wall looked seamless. One wall partition after another marched across the front of his cube, with no hint where the opening had been. He wouldn't even know where to stop!

She was pulled out of her reverie by the distant click of a door. Oscar had arrived.

She slouched low in her chair, listening to the soft rhythmic crush of footsteps on the carpet as he strode past the cafeteria and into the developers' area.

Then silence. Her smile widened and she waited for the explosion.

Oscar inhaled audibly and bellowed, "WHERE . . . is my OFFICE?!?"

She snickered in silence; Oscar refused to refer to his workspace as a cubicle.

"Is this the appreciation I get for my vast technical knowledge, and for sharing my deepest development secrets to help mere developers become renowned embedded systems wizards?

"And what, specifically, am I being thanked *for*? For helping Josie solve the amazing wandering labeling machine problem? Or perhaps it was Ravi, who tracked a grinding motor from hardware to software back to hardware for two days until he realized it was an initialization problem? And it most *certainly* cannot be for helping Li Mei with her erratic tape measure problem, because I don't think she has been here long enough to be corrupted by the likes of two other instigators, who shall remain nameless!"

Silence.

"But, perhaps," he drawled, "I could be mistaken."

Josie could almost feel Li Mei cringe.

And, in true Oscar fashion, he did what no one expected.

After gently leaning his bag against the wall, he strode off. A few minutes later, he reappeared with a 6-foot step ladder, which he opened and straddled over his new wall. Without delay, he snagged his bag on the way up and then descended into his cubicle without another word.

Josie dropped her head into her hands, chagrined.

A new day had begun.

While Josie was taking the heat from Oscar for their morning prank, Li Mei slipped out to the lab to review everything they'd discussed the day before. In her clean handwriting, she listed everything on the whiteboard.

Knowns:
- LED blinks at same rate as actual heart rate.
- LED based on hardware signal, so gives no information about software.
- Displayed heart rate about twice actual heart rate.
- Displayed heart rate tracks changes in actual heart rate.
- ECG waveform in Austin looks realistic, just too fast.

She added some questions she should have investigated before bothering Josie.

- Valid ECG signal, properly conditioned, is available at the A/D converter input pin - is this true?
- Software is configured to collect A/D data every 1 msec - is this true?
- Software sends data to Austin every 1 msec - is this true?

Her gut told her she had an off-by-two problem in the software. If she could answer these questions, that might help her locate the bug.

Narrowing down the problem was easier when she applied Josie's logic. She admired Josie's thought processes when they walked through a problem or brainstormed a new idea. Josie tended to think out loud, debating ideas with herself while Li Mei watched.

It was a good way to learn a lot in a short period of time!

She set to work connecting the digital oscilloscope to the input of the ADC. After she reconnected her ECG leads, the ECG began marching across the display with its rhythmic peaks shooting vertically to fill the screen. The time intervals convinced her that the ADC was seeing the correct ECG analog signal. That short test answered the first question and had only taken 30 or 40 seconds.

Next, she turned her attention to the debugger and decided to restart from scratch, in case the debugger was in some sort of unhappy state.

In fact . . . she stopped in midthought and peered back at the circuit board. Maybe she *should* change the hardware so the micro turns the LED on and off to save components. She would ask Tom about that later.

With the debugger loaded and ready to run, she verified that she had enabled only one 12-bit ADC and was using the TIMER function to periodically initiate conversions every 1 millisecond. Each time the TIMER interrupt fired, she reloaded the timer register with a number she thought would produce a 1-millisecond sampling interval. Time to check if the reload was wrong.

Verifying the sampling frequency shouldn't be too hard, and she remembered how Ravi had measured execution time on the Animal project. Since her main routine waited for a flag that was set in the TIMER to indicate new data was ready, she could measure the time between consecutive write accesses to this variable! With a plan in mind, Li Mei set her breakpoint in the main routine on the line of code that stored the latest ADC acquisition, and executed the program.

$$BKP \rightarrow new\_ECG = New\_ADC\_result;$$

Almost immediately the debugger stopped with the program counter on the breakpoint. The trace buffer showed timestamps of all commands executed right up to the breakpoint, but as she scrolled back through the buffer, she couldn't find the previous execution of that line.

"Stupid." She realized that the breakpoint stopped program execution the very first time the line was reached; she didn't have a previous timestamp to measure from. She crossed her arms on the bench and pondered how to make the program skip the first instance and stop the second time.

A new CD started playing in the hardware area; she hadn't realized it had gotten quiet. She let the words flow over her and hummed along.

"Oh, stupid again! I don't need to set any breakpoints. I just let it run and grab two successive timestamps." After several ECG cycles went by on the scope, she stopped the program and scrolled backwards through the trace buffer, finding four times when new data was acquired from the ADC. She quickly performed the subtractions and determined that each time interval was almost exactly 1 millesecond. Then she remembered she could have configured the debugger to capture timestamps of several variable write accesses in a row, but she already had her answer.

The ADC subsystem was working correctly.

"Hey, I'm sorry. I got caught up with Oscar."

Josie had walked up between the row of lab benches and dropped onto a stool next to Li Mei. "He made me use the ladder. I think he's going to leave it there."

"Really? Is he mad?" Li Mei looked worried again.

"I'm not sure. He didn't say much; he's been working really late lately so maybe he's just tired." She shrugged. Not much she could do now. "But I accepted complete responsibility. So, progress?"

"Well, I logically wrote down what we know, and now I am answering some questions I came up with. I verified the hardware is okay, and that I am collecting data every millisecond. The next test is to verify that I am sending it to you every millisecond."

Josie stopped what she was about to say and looked at Li Mei, thinking about her own design. *Every millisecond? She's sending me data every millisecond?* It was possible her architecture could support that, but she hadn't anticipated a messaging rate that high.

"Okay, tell me about your main loop and what data you are sending me. Is this the flowchart?" Josie picked up the paper laying on the bench.

"Yes. I verified this **new_ECG** variable is written with new data every millisecond. Then I take that variable and construct a message with that variable and then send it to you. Then I reset the flag to tell the ADC I can receive more data."

"What does the flag do?"

"I set the flag in **TIMER()** when I get new data and then reset it in the main loop. I learned from Ravi not to put everything in the interrupt and he suggested using a flag instead."

"But does anyone look at that flag?"

"Oh, yes, I am sorry - in the main loop I only send you new data when the flag is set."

Josie chewed on her lip, wondering what was happening. Li Mei was waiting for a send acknowledgement from the Austin Monitor before she reset the flag, and the ACK was probably taking too long.

"Do you know what the maximum data rate is for this wireless protocol? Does it support the data rate you are assuming? And, by the way, what data rate ARE you assuming?"

Li Mei fidgeted in her seat but remained silent.

"Why am I asking these questions?"

"Um. If it takes longer than 1 millisecond to send the data, then Austin will not receive all data."

Damn, she was catching on quick! Josie wanted to smile but was torn between being a teacher and a teammate.

"And how often does Austin receive the data?"

"About once every two milliseconds. I was thinking there was a factor-of-two problem." Li Mei looked at her with eyes dark and entreated, "That was my hypothesis - I already thought of that! I was just testing my questions to be thorough first!"

"Okay, then explain exactly what is going wrong."

Li Mei composed herself again.

"Okay, I think it works like this." She settled her notepad on her knees and drew a quick sketch with narrow pulses representing each ECG sample collected by the TIMER. (Figure 7-4.) "I collect a valid ECG sample every millisecond and set the flag every millisecond." She drew dark vertical arrows to emphasize setting the flag.

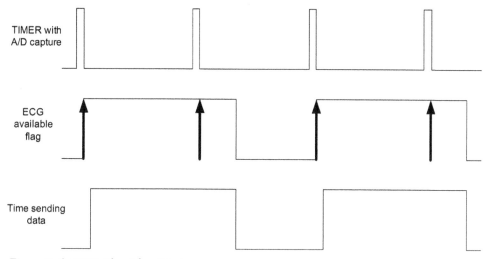

Figure 7-4 ECG Algorithm Timing.

"If I initiate a data send to Austin and wait for an acknowledgement, it is possible that another TIMER interrupt happens during that time." She drew a second dark arrow. "I ignore that data because I am still working on the last data value."

She paused and drew wide pulses representing the acknowledgement time. "After Austin replies back, I automatically reset the flag, even though that second new data is available."

She looked back up at Josie with discomfort. "This software purposely ignores every other value."

Josie nodded. "**TIMER()** can read data and set the flag in the interrupt, but it's your responsibility to grab the new data point before the timer goes off again and overwrites the flag. And you're not storing a timestamp, so you have no way to recover from a missing data point."

Li Mei slumped down over her stool. Josie realized that, although Li Mei was smart, her newbie insecurities were sometimes close to the surface.

"Hey - we can fix this. But you didn't tell me what data rate you are assuming."

Li Mei hesitantly picked up her calculator but appeared unsure where to start.

"You're sending one sample - two bytes of data - every 1 millisecond, right?"

Li Mei nodded.

"That means each message sent to Austin is really 15 bytes long."

She saw Li Mei's eyes widen in surprise. "Yup. Welcome to the real world. Every message contains about 12 bytes of overhead. That's the message header, the sequence number, and some other address and checksum information that the hardware physical layer uses to transmit the data. And then there's one more byte for the length of the payload. That's a total of 15 bytes just to send your two bytes of data."

"Then my data rate that I *thought* I was sending is . . ." Li Mei bent over the calculator in earnest. ". . . 117.2 kilobits per second. I got that by dividing the sampling time interval of 0.001 second into 15 bytes, and then converting to kbps." She showed Josie her pad. "Is this right?"

$$\frac{15 \; \text{bytes}}{0.001\text{s}} \times \frac{8 \; \text{bits}}{1 \; \text{byte}} \times \frac{1\,\text{kb}}{1024 \; \text{bits}} = 117.2 \; \text{kbps}$$

Josie looked over at the pad, and then shook her head.

"Almost. You're thinking about the "kilo" conversion with respect to bytes and data storage. In that case, 1 kilobyte is equal to 1024 bytes. With data transmission, "kilo" means one thousand, and a kilobit is 1000 bits."

"That's strange." Li Mei paused. "So you have to know if someone is talking about computers or telecommunications to know which they mean? Hmmm." She corrected her equation and then got a nod from Josie.

$$\frac{15 \; \text{bytes}}{0.001\text{s}} \times \frac{8 \; \text{bits}}{1 \; \text{byte}} \times \frac{1\,\text{kb}}{1000 \; \text{bits}} = 120 \; \text{kbps}$$

"A data rate of 120 kilobits per second is actually pretty fast, but acceptable, for this wireless protocol. This protocol can't compete with higher data rate solutions, but it has incredible battery life, which means we don't have to change the sensor batteries for a long, long time."

Josie leaned back and pulled her hair into a ponytail. "Now, Ravi had a similar problem with an interrupt and some processing that didn't finish in time. How is this different from his problem?"

"Ravi's interrupt took too long, but the long ones only happened every 30th cycle. He moved the long functionality out, but that functionality still happens as often as the hardware needs." After a moment of thought, she continued. "Mine requires long functionality every single time. There's no way to just move it out of the TIMER. It will take too long no matter *where* I put it in the software."

She appealed, "What do I do now? I didn't think it would take that long to send the data. The specification requires sampling at 1-millisecond intervals."

Reader Instructions: Without knowing anything else about the two systems, what can you suggest as possible solutions to Li Mei's problem?

"Well, the first thing to remember with embedded systems is that you usually have much lower processing power than with a personal computer. Your data method would work on your desktop but these micros operate at a slower speed in order to conserve battery life. Make sense?" Josie looked her in the eye to see if she understood the issue.

"Yeah. I guess so. So can I fix it by resetting the flag before I send data to you, instead of after sending?"

"That's a thought. That could prevent **TIMER()** from setting the flag when it's not yet reset, but it doesn't stop the main problem of taking too long to send the data and receive the acknowledgement in the first place." Josie shifted on the stool.

"Instead, I'd like to suggest something a little different. Tell me what the biggest time bottleneck in this system probably is."

"I think sending you data," Li Mei admitted.

"Probably," Josie nodded. "Communications between processors requires synchronized timing or handshaking of some sort. One usually has to wait on the other for a task to be complete. You've stopped everything in your system except the TIMER to wait for my acknowledgement for each piece of ECG data. But do you think I need each one sent to me in real time?"

"The requirements say that real time must be maintained, so I wanted to make sure you had the data as soon as I collected it."

"Ahhh, I understand the confusion." Josie readjusted the stool again. "You are taking the requirements literally, which, don't get me wrong, is usually a Good Thing. In this case, they are a little vague. One millisecond is the required sampling rate of the final stored data, but the "real time" comment refers to my numeric display of heart rate in beats per minute. I should've clarified that.

"I update the numeric value every second or so, but not every millisecond. But since heart beats are rather slow in the grand scheme of things, it isn't critical if I'm late by a couple hundred milliseconds. The old value stays on the monitor and heart rate doesn't change by a huge amount beat-to-beat. So that means, on your end, you don't need to send me data every millisecond. You can save up a set of data values and send a squirt of them at once."

"A squirt?" Li Mei looked doubtful.

"Well, like an array of them. Maybe 10 at a time. Then I read and store them, and make any adjustments to the real-time display and the end user never knows." She added, "We should have reviewed your design concepts before now."

Li Mei stood and motioned her to stop. "I know. I like to figure things out on my own, and I am not really done with the design or anything. I am just learning the communications protocol and didn't make anything in stone yet."

"That's fine. Personally, I like to muck around with some ideas, and then write some sample code to help me think about how the final design should work. The first code you write shouldn't be what ships anyway. I'm surprised that you got the two processors talking to one another without any help. What message format are you using to send me the data?"

"Just like Message ID number 6, but with the ECG_DEVICE message ID you gave me. I looked at your interface document and found a simple message for one data at a time so I wanted to try this one and make things work."

Protocol Packet Format

Header	Address	Message ID	Group ID	Data Length	Data	CRC
4 bytes	4 bytes	1 byte	1 byte	1 byte	0-27 bytes	2 bytes

Message Name	Message ID	Data Length	Data Format	
Raw Data Message Formats				
DATA_CHAR	5	1	Single byte	
DATA_INT	6	2	1 2-byte integer (high byte, low byte)	
DATA_LONG	7	4	1 4-byte long (high byte first)	
DATA_FLOAT	8	4	1 4-byte float (IEEE format)	
DATA_DOUBLE	9	8	1 8-byte double (IEEE format)	
DATA_CHAR_ARRAY	10	1 + (0 - n)	Num chars (1 byte)	Raw char data
DATA_INT_ARRAY	11	1 + (0 - n)	Num ints (1 byte)	Raw int data
DATA_LONG_ARRAY	12	1 + (0 - n)	Num longs (1 byte)	Raw long data
DATA_FLOAT_ARRAY	13	1 + (0 - n)	Num floats (1 byte)	Raw float data
DATA_DOUBLE_ARRAY	14	1 + (0 - n)	Num doubles (1 byte)	Raw double data
Formatted Data Message Formats				
PULSE_OX	20	6	Timestamp, pulse-ox data (4 chars, 1 int)	
BLOOD_PRESSURE	21	24	Timestamp, 10 samples BP data (4 chars, 10 ints)	
FOOT_BEND1	22	25	5 Bend Groups. Each is 5 bytes (4 10-bit INTs, bit-packed)	
ECG_DEVICE				

Figure 7-5 Austin-to-Sensors Communications Protocol Data Packet Formats.

Josie held out her hand and Li Mei passed her the document. (See Figure 7-5 [3].) "Oh, yeah. Hey, remember those message overhead fields I talked about? Here they are and you can see where the extra 12 bytes comes from."

"I wasn't sure what that meant," Li Mei admitted.

"All of those bytes get sent with your data," Josie told her while looking at the sheet. "I have message IDs for sending individual chars, ints, longs, and doubles, but they're not the best for data streams. I assumed you'd do something like the formatted messages at the bottom."

She handed the sheet back to Li Mei. "Take a look at Message ID 21 for blood pressure; it sends 10 samples together with a timestamp so we know the sampling interval and can detect if data are missing. Model yours after that one, and keep in mind that the max data payload for any message is 27 bytes."

Li Mei accepted the sheet back and made a small star next to ID 21.

"So if I asked you to change your design to send me several samples at once, would this solve your problem?"

"I think this could help. But now I am confused a different way. Why don't I calculate the heart rate directly and just send that to you instead of all the raw data?"

"If Austin was just a data collector and router, that would be ideal. But Ravi's signal-processing analysis needs your raw data at 1 kilohertz for the advanced diagnostics, like heart rate variability and prediction of disease state."

"Oh yeah, I forgot." She added with a knowing smile, "And that means 1000 Hz, not 1024 Hz, right?"

"You got it." Josie laughed as she stood up. "Come get me when you've done another round of analysis on your design?"

"Yes, I will. Thank you, Josie."

Another doubt bubbled up and Li Mei turned back. "Don't you have to rewrite the Austin software depending on how I change the ECG message format?"

Josie broke into a grin. "That's what's cool about this design - since your message format contains the length of data sent, my receiving software can be more flexible. If you send five samples, I store five. If you send eight, I store eight. However, my ECG software assumes payloads larger than 5 bytes contain a timestamp in the beginning. Since your last attempt only sent one sample in the message, I assumed no timestamp and used the default sampling rate of 1 kHz."

"Oh! That is a mystery solved!" Li Mei had wondered how Austin had translated her messages the first time. It was amazing how the software magically knew what she had sent; what a robust system.

Reader Instructions: Before looking at Li Mei's redesign, update her design yourself. A) What should the new packet format look like? B) What changes to her flowchart are needed in the TIMER interrupt and the main loop? C) Based on your proposed changes, how often will each message be sent? D) Using the Protocol Packet Format in Figure 7-5, compute your new data rate in kilobits per second (kbps).

The next morning Li Mei burst into Josie's cube.

"Josie! It's working - come on and see!"

Li Mei led the way back to the lab, this time in much better spirits. Josie thought she would be annoyed being interrupted this often, but she liked Li Mei's drive and transparent excitement when things worked. Li Mei was fun to work with.

"Okay." Li Mei stood beside her bench. "I changed my main loop. I have a new array to hold 11 ECG values at a time. Each time new data is acquired, I save the value to a new array and reset the new ECG flag." As she explained, she reconnected her ECG leads to the breadboard. "After I collect all 11 values, I construct a message to you using a new message format with a timestamp and I send it."

Li Mei consulted her notepad. "Also, I checked several things before bothering you. First, I found that 11 values are stored and they are consecutive values and no values are lost. Next, I found that I can store all 11 values in one message and send it. And, finally, I found that your display will say I have a heart rate of 76 beats per minute."

Li Mei stood both expectantly and triumphantly tethered to her lab bench by the ECG wires that trailed under her blouse, grinning at Josie.

"See, it's working!"

Josie checked the display and the debugger, and grinned widely at her. "Sounds like you have it all under control. Care to hop on the Habitrail again?"

Li Mei immediately began jogging in place. As expected, the heart-rate display climbed from 76 BMP to over 100 BPM.

"It looks like it's working, Li Mei. How did you test that you aren't missing any samples?"

"I created two buffers; each can store 11 values. Since sampling occurs every millisecond, it takes 11 milliseconds to fill a buffer. When one gets full, I send it and start saving data in the other one, and toggle between them every 11 milliseconds. After I send one array of 11 to Austin in the main routine, I check for communications back from Austin. When Austin acknowledges the message, I clear that array. At that time, the TIMER should still be writing to the other array." Li Mei looked at her. She was still jogging. "Do you understand so far? It is in my new flowchart (Figure 7-6)."

"Yes, I get it." Josie nodded and motioned for her to stop. "In TIMER, you check when you are ready to use that new buffer, and if it hasn't been cleared out, you know you are about to overwrite data."

"I set a breakpoint on that line and let the program run for almost an hour and the debugger never stopped. That means no samples were lost."

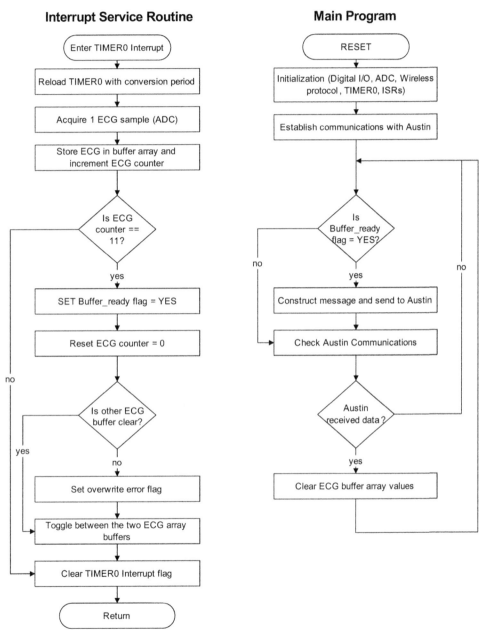

Figure 7-6 Updated ECG Sensor Software Flowchart.

Josie bit her lip and debated with herself the best way to approach this change. Li Mei hadn't removed the problem; she had just made it far less likely to happen. But would this be enough? She explained her concern to Li Mei, while watching her face drop in disappointment.

"You've got the right idea - you need to buffer your data. But what happens if I put the Austin Monitor in the other room and it has trouble receiving your data correctly?"

Li Mei quickly pointed out that the software had resend capabilities.

"But what if the resend takes longer than 11 milliseconds?"

"Oh. Then I have an overwrite condition again." Li Mei dropped to the lab stool and looked forlornly at the ECG circuit board. After several moments she offered, "I could add a few more buffers, but that doesn't fix the problem, does it?"

Josie shrugged. "I don't know. Take a look at the requirements to understand what the maximum error rate can be. For my pulse oximetry, missing a few samples isn't a problem, but the requirements for the ECG data analysis might be more stringent."

Later in the day, Josie worked her way back to the lab to check on Li Mei. She had seen her and Ravi discussing the requirements for the ECG analysis over lunch, and Ravi had shown her some of the signal processing he had performed on her sample data. From their laughter and the closeness of their chairs, it looked like Li Mei felt comfortable working with Ravi.

The team's dynamics had gelled a little differently than she had expected; Ravi always seemed to butt heads with Oscar for reasons she didn't quite understand, but his comfortable rapport with Li Mei kept the group from slipping into dysfunctionality. She had thought, and even hoped, he might come to her more for help, but he rarely did.

Josie spent a good part of her lunch hour in her cube thinking about Li Mei's design and its implications for her own part of the project. The Austin Monitor could receive a large volume of data, and she had tested her received-data storage algorithms at the maximum wireless data rate successfully.

The ECG was a critical signal. In addition to heart rate, the customer also wanted to understand how small changes in heart rate could be used to predict a disease state. If they could perform the analysis on small segments of the data or could interpolate any missing data on the receiving end, Li Mei would have an easier time with her design concept. But if the algorithms required an hours' worth of data with absolutely no missing samples, her current design concept would require major revisions before it would pass a final review.

She concluded that, however Li Mei managed to send the data, as long as it used the ECG_DEVICE ID and the data-length variable matched the actual payload, she was sure Austin would be able to extract it and store it.

Back in the lab, she saw Li Mei sitting at her bench. Li Mei was intently focused over the Austin Monitor, and she didn't see Josie until she dragged a stool over to join her.

"Hi, Li Mei. What did you find out about the requirements?"

"Oh, hi, Josie. I am running a test now where I walk in and out of the lab so my sensors drop communication with Austin, and I am checking to see if I lose any data." She was quick to add, "It is not a formal test, though."

"That's okay. It's a start."

"The customer is using different algorithms for the analysis, but they are looking at 5-minute measurement intervals of ECG data, and they will scan through the files for good intervals. They also focus on the time interval between two R waves, so missed samples at other places in the waveform are okay. And at our sampling rate of 1000 Hz, interpolation can be used to correctly isolate the R wave point."

"So did you increase the number of buffers?"

"Yes. I am storing the timestamp with each buffer, and also a flag I check to see if it has been acknowledged by Austin before I erase it." Li Mei explained that, even though each buffer was small, she was limited to less than a minute of wireless signal loss before she was forced to overwrite previously sent but unacknowledged data. The inexpensive microcontroller didn't have a lot of memory.

Josie agreed with her assessment. "Another disadvantage of these small portable devices is power consumption. Have you thought about how sending several samples at a time in one message will affect that? And how did you come up with 11 samples anyway?"

Li Mei pulled out the interface document for the wireless protocol they were using for the monitor and opened it to a page she had marked with a yellow sticky note.

"My new ECG message format is based on the blood pressure formats. The payload is 26 bytes - I can squeeze in 11 two-byte ECG values and one four-byte timestamp. Each ECG is two bytes. I thought about bit-packing like you are doing, but that may take too much time to pack and unpack, and it only allows me to send another three values per message."

"Okay. If we need to bit-pack we can do that later. Sending me 11 data points every 11 milliseconds is a lot more reasonable than sending me 1 data value every millisecond! Let's figure out how much overhead you saved."

Josie reached under the shelves to her bench and pulled out a notepad. "I remember your first method required 15 bytes per message. Multiply that by 11 data points and you get a total of 165 bytes sent over the air every 11 milliseconds. Using your new method to send eleven in one message, you only send, ah,

12 bytes of overhead plus 1 length byte plus 26 bytes for the data and timestamp. That's 39 bytes for the same 11-millisecond interval."

Li Mei nodded. "That's a lot. Over four times less data. I hadn't thought about all the overhead bytes the first time."

"Yup. That extra 126 bytes might not seem like much to worry about, but what happens if someone wears this monitor for several hours?" She punched some values into Li Mei's calculator. "For a three-hour trial, you'll send roughly 153 MB compared to 37 MB. You learned from the trade show that power consumption is larger when the processor is transmitting because it has to spew all those bits in the air. Total power consumption is about 10 times greater with the transmitter on, so a major goal is to keep that transmitter off as long as possible."

"Wow. So just changing the messaging will make the batteries last a lot longer." Li Mei smiled and grabbed her calculator. "That means my original data rate of 120 kbps is now about 29 kbps."

$$\frac{39 \; \text{bytes}}{0.011s} \times \frac{8 \; \text{bits}}{1 \; \text{byte}} \times \frac{1\,kb}{1000 \; \text{bits}} = 28.4 \; kbps$$

"That's a lot better." Josie changed gears. "Your design idea seems reasonable and it looks like it will satisfy the requirements. Document all of this and we'll have a design review."

"Wait, I'm not sure about something. How come I just finished the software, and *now* we are having the design review? Isn't that out of order?"

"Don't jump ahead of yourself; you're not done with the software."

Li Mei stared back at her, hard. "I'm not?"

"What you did is a design concept. You explored several design ideas to satisfy the requirements, and you settled on one design that seems to be best, right?" Josie paused, but Li Mei continued to look confused. "Now you need to document the design you are proposing and the reviewers will make sure it really does satisfy the requirements. Technically, you don't write the code until after that."

"But if I wasn't supposed to write the code, then how was I supposed to choose the best design concept?" she challenged.

"I know, it seems like a catch-22. But now you have the opportunity to go back and start over. Code can get kludgy after several design changes; you get sloppy about variable names, forget to delete code you aren't using anymore, stuff like that. And then you have to consider things like error checking, testability, and maintainability. Also coding standard stuff and comments." She paused. "Did you do all of that already?"

Li Mei exhaled deeply in defeat and admitted that she hadn't.

Josie returned her calculator. "Don't worry. We don't get to do the normal design process steps often enough, and when we do, it seems strange, huh?"

"Yeah, but I guess it makes sense. The mechanical folks do drawings and make models, and the electrical engineers use circuit simulators to develop their designs. Maybe just we software people are strange that our prototyping method isn't different from our final deliverable."

"That's a very good point." Josie stopped to think about Li Mei's assessment. "I prototype a lot with flowcharts. Too bad there isn't such a thing as a flowchart test tool that checks the logic of your flowchart!"

After sharing a laugh, Josie asked, "Now tell me, what are the take-home messages for this part of the project?" As Li Mei reached for her lab notebook, Josie realized that she had anticipated this question. It was becoming a habit.

Li Mei began, "I have several things. First, transmitting too much data drains the battery quickly, so reduce overhead by combining and compressing data if possible. Second, using a flag in an interrupt or timer function can be dangerous if the main routine is polling the value of the flag. You can miss samples. Third, controlling the LED from the microprocessor saves cost of parts and shows that the ADC subsystem is active, although a trade-off is we lose one digital output pin."

She paused, trying to remember anything else to add to the list. "Oh, yes, and don't be afraid to do exercising to debug your code!"

Oscar entered the lab and walked matter-of-factly up to the group before stopping to lock his hands behind his back. He dropped his head momentarily before looking up sharply to address them all.

"I am placed in the decidedly awkward situation of rewarding you for your bad behavior of yesterday. Randy, of all people, actually negotiated that some of the pie promised by B&L Telecom be trickled down to the engineers who made the World Wide Telecom Austin demo a success."

Josie looked expectantly at Ravi and Li Mei, and then turned her attention back to Oscar.

"I've gotten approval for the entire team to go to the Embedded Systems Conference next week."

"Cool!" Josie exclaimed. "This is great!"

Oscar allowed a high five with her, while Ravi and Li Mei looked on in anticipation. Oscar noticed their confusion and went on to describe the conference as a vital resource for embedded engineers for tools, products, and educational courses and seminars.

"I've been lobbying for this for a while, and with the team's performance in making that trade show a success, it was easy for me to convince Randy and upper management that taking you to the pinnacle conference for our field was just as vital as sending the marketing folks off to telecom industry trade shows and golf outings."

With the announcement fully conveyed, he finally broke a smile and congratulated them for the well-deserved reward. Josie watched the smiles fill everyone's face - appreciation that their work had finally been recognized.

"If you can, talk to Kathy by tomorrow and she'll make the reservations. We leave on Wednesday."

The conference was an overwhelming success and the team excitedly traded notes over dinner as they waited for their cancelled flight to be rescheduled. Josie hoped they would be able to make the conference a regular event; the three days had gone by too quickly.

Pausing with sandwich in hand, Li Mei answered her question.

"My favorite parts of the conference were the classes about programming like *Embedded Systems 101* and *Understanding Interrupts and Priorities*. I learned new things about complicated systems with real-time operating systems. And there were so many people in the lectures, both younger and older than I am."

As she paused to take a bite, Ravi interrupted, "What about the huge leather lounge chairs, can you believe that?" An incredulous grin lit his face. "Right in the middle of the show floor booth. Just for us to sit there and watch product demos!"

Oscar nodded. "It's one of the few places where embedded folks are treated with a great deal of respect - the show is for us and they cater to us. They listen to our ideas and complaints about their products and they offer in-booth demonstrations for nearly anything from chip evaluation kits to full suite development tools." He scooted his chair in to let people pass behind him in the crowded restaurant. "I'm happy you attended so many of the demos, Ravi, but I hope it was for the education rather than the comfort."

"No," Ravi tried to look abashed, "but I did learn a lot. Li Mei and I went to a lecture on reducing power consumption in portable devices. They introduced some energy-saving techniques in hardware and software like using idle and sleep modes to turn off parts of the microprocessor that aren't being used, and scaling the clock frequency. They even gave examples of software that performed the same functions but used different amounts of power. They also showed battery voltage and current drain plots for a cell phone, an implantable defibrillator, and a digital camera."

Josie didn't remember reading about that lecture, and she leaned over the table as Ravi located a pen and a clean napkin. He drew several jagged lines that looked like rectangular waves stacked on top of one another, making deep peaks and valleys. After he labeled some of the valleys TX ON, she realized he had drawn the battery voltage curve.

He pointed to the valleys. "Look at how much a high current drain event like turning on the transmitter affects the instantaneous battery voltage. You can tell from the battery-voltage plot exactly when the transmitter is on, when the display backlight is on, and you can even see little steps when the phone is searching for a new system."

Li Mei added, "It's like you can see the battery gauge falling! Now we understand how hard the phone was working to find a good system at the trade show, and why it got so warm."

"It would be a good exercise for you and Josie to do the same analysis with Austin when we get back," Oscar said. "Run the battery-voltage signal into a spare A/D input on the microcontroller."

"Yeah, we can do that." Josie reached over to snag the ketchup for her fries. She liked Oscar's idea; they might find other parts of the code that caused higher battery drain they weren't aware of.

"The vendors had a lot of good chotchkas, too. I got CD-ROMs and pens, but did anyone get the microcontroller-based name tag that displays your name and plays games? That's a great idea for an inexpensive evaluation board."

"So, I am imagining you next to Ravi in those easy chairs. You didn't learn anything either?" Oscar raised his eyebrow at Josie while wiping his mouth.

"No, I attended every lecture on embedded-systems disasters and I learned how simple coding bugs that could have been caught with a code inspection ended up killing people."

Li Mei leaned over, "I think he's kidding, Josie."

"I know." She waved an apology; she hadn't meant it to come out that way. "The reality-based lectures were awesome; it's just scary how easy it is to ship a product with defects. Those stories keep coming back and I think about times I don't test my code as well as I should. I'm going to go back over the Austin test plan to make sure we've covered all the bases on the error-checking logic."

"Your test plan was fine," Oscar assured her. "Put that energy into helping Li Mei develop a good one for the ECG sensors."

"I'll do that." She nodded and then asked him, "So what was your favorite part?"

"Hmmm. That's tough." He tipped his head back to think. "Aside from the T-shirts to add to my extensive collection of free vendor-wear, perhaps the open forums at breakfast and lunch, like the ones on wireless security and medical applications. They had a facilitator and expert in the room to provide the cutting-edge information, but the forums allowed everyone to talk about feasibility, cost, and implementation time.

"Also, not to be politically incorrect," he added, "I was pleased to see more and more women joining the field and showing up at these conferences. This is no longer a male-dominated field."

Josie raised her glass. "That's right. And we'd like a little more gender-equity in the musical selections in the lab, too. Right, Li Mei?"

"Oh, I don't care," Li Mei said innocently. "I just tell Ravi I'll email his girl-friend about the pretty movie star on his computer and he lets me play whatever music I want."

Additions to Li Mei's List of Debugging Secrets
Specific Symptoms and Bugs
- Proportional errors (*twice as fast, off by 3, etc.*) may mean data is periodically missed or two entities expect data at different rates. Or, just a bad configuration setting.
- Be careful using flags for signaling when two different entities set and clear the flag without checking its state first.
- Reduce overhead when transmitting data to conserve battery power. Sending more than one data sample at a time and/or compressing data in the message payload can radically increase the battery life of wireless devices.

General Guidelines
- Don't make assumptions about how something was implemented (e.g., in hardware versus in software). Look at the documentation.
- Don't assume commercial tools are bug-free; if a tool does something strange, suspect the tool.
- When you feel overwhelmed, take things one step at a time.

Chapter Summary: The Case of the Rapid Heartbeat (Difficulty Level: Easier)

In this mystery, Li Mei sends real-time ECG data over the air to the Austin Home Medical Monitor, but finds her own heart rate reported at nearly twice the actual rate. She and Josie find that the data collection and data transmission rates are not the same because a flag to communicate between the periodic TIMER and the main loop causes every other sample to be lost. Her redesign includes buffering the ECG samples and reducing transmission overhead by sending several samples together.

The Problem Symptom(s):

- Li Mei's true resting heart rate (90 BPM) was displayed on the Austin Monitor as 180 BPM.
- The displayed heart rate tracked changes in her own heart rate; it increased when she jogged in place.
- The LED blinked with her true heart rate.

Targeted Search:

- Verified that the LED circuitry is hardware-based and not related to the symptoms.
- Verified that unplugging all other sensors from the Austin Monitor does not affect the symptoms.
- Found that raw data received by the Austin Monitor revealed a reasonable-looking ECG signal with a frequency of twice Li Mei's heart rate.

The Smoking Gun:

The device collected a new ECG sample every millisecond, but only sent data to the Austin Monitor every other millisecond.

The Bug:

Every other ECG sample was ignored.

- A global variable flag was used to communicate between the TIMER interrupt and the main loop. It was set in TIMER when an ECG data point was acquired and cleared in main() when an acknowledgement was received from the Austin Monitor. When the Send-ACK cycle took longer than 1 millisecond, the main loop reset the flag, causing a subsequently acquired data point to be ignored.

The Debugging Method Used:

- Verifying the A/D sampling rate and Austin Monitor receive data rates.
- Hypothesizing the bug based on the evidence.
- Comparing symptoms to previous projects.

The Fix:

After verifying the sampling requirements, three changes were made:

- Transmitted 11 ECG samples in each message rather than 1. This has a positive side effect of decreased battery drain and overhead.
- Created several buffers that can store almost one minute of data for temporary loss of the wireless link.
- Added a timestamp to each message to allow detection of missing data and interpolation on the receiving side.

Verifying the Fix:

The design concept was evaluated prior to the design review:

- Using timestamps and monitoring a data overwrite flag, several hours of error-free data transmission were verified.
- No data loss with short wireless connectivity failures was found.

Lessons Learned:

- Communications between synchronous and asynchronous processes must be carefully designed and tested.

Code Review:

This chapter presented no software, but the flowcharts revealed the architecture problems explored above.

What Caused the Real-World Bug? During the investigation, Mitsubishi Fuso Truck & Bus Corporation found that some of their commuter busses were susceptible to high-powered radio-frequency broadcasts. They announced that the braking systems may not work properly within close range of high-powered electromagnetic interference (EMI). In the accident cases, the hardware and software braking system designed to detect a wheel-lock condition while busses were underway was suddenly triggered. Software stopped applying the brakes, effectively preventing the drivers from bringing the bus to a stop!

The source of the interference? Nearby high-powered (and quite illegal but common) radio signals. Some trucks modify their CBs to pump out 1000 watts or more - significantly over the 4-watt limit.

The moral is to expect the unexpected. The real world is rife with noise, electromagnetic interference, and signals from all sorts of sources. Hospitals are particularly problematic as electronics, all spewing some level of radio frequency interference (RFI), get packed into every spare corner. [4]

References

[1] Poll: How much time do you spend debugging your embedded system? *Embedded Systems Design* magazine, *http://www.embedded.com/pollArchive/ ?surveyno=5900061.*

[2] Goldberger AL, Amaral LAN, Glass L, Hausdorff JM, Ivanov PCh, Mark RG, Mietus JE, Moody GB, Peng CK, Stanley HE, "PhysioBank, PhysioToolkit, and PhysioNet: Components of a New Research Resource for Complex Physiologic Signals," *Circulation* 101(23):e215-e220 [Circulation Electronic Pages; *http://circ.ahajournals.org/cgi/content/full/101/23/e215*]; June 13, 2000.

[3] TinyOS, an open source operating system designed for wireless embedded sensor networks, *www.tinyos.net.*

[4] Ishikawa C., "Loss of bus braking due to nearby illegally modified transceivers," *The Risks Digest: Forum on Risks to the Public in Computers and Related Systems,* 2003, Volume 23, Issue 9.

Additional Reading

Embedded Systems Conferences, Connecting Engineers & Developers with the Practical Skills & the Latest Technologies, *http://www.esconline.com/.*

Task Force of The European Society of Cardiology and The North American Society of Pacing and Electrophysiology, Heart rate variability: Standards of measurement, physiological interpretation, and clinical use," *European Heart Journal* (1996) 17, 354-381.

Real-World Bug [Location: Somewhere, USA] A technician was repairing a dusty old instrument, years after it had been originally programmed and sold. As he fiddled with it, the 7-segment LEDs started flashing "HELP HELP," confirming his long-held belief in the supernatural. He rushed, ashen-faced, to the nearest embedded-systems guru for deliverance.

Chapter **8**

What Kind of Error Message is "lume Fault"? When Symptoms Seem Impossible

Mike walked into the development office area, trying to remember where Josie's cube was. She hadn't been in the lab all day, but her email .sig mentioned Feynman Lane so he decided to pay her a visit. Passing Edison Avenue, he recalled when they'd renamed the aisles after engineers and scientists. He remembered rolling his eyes, shaking his head in amusement to see all the geeks hanging over the cube walls with an arsenal of hardware installing the signs.

But, he grudgingly admitted, it *did* help with directions.

After dipping his head left and right up the long Feynman Lane aisle, he finally spotted Josie's long ponytail in an amazingly neat cube festooned with Far Side and Dilbert cartoons.

"Heyyyy, Josie. Got a minute?"

She looked up and gave him a smile, motioning to her guest chair. "Hi, Mike, what's up?"

"Tried finding you in the lab - you've been scarce lately."

"I'm just finishing up the code review and documentation for the Austin Monitor." She motioned to a stack of code listings and printouts on the floor next to her chair. "That was a fun product, but I'm glad it's finally being delivered to beta testing. All the severity one, two, and three bugs are fixed, but I'm sure more will crop up in beta."

Listening, he looked nonchalantly around her cube and spotted a small stuffed ladybug perched on her computer and a circuit board pinned to the wall with "Good Luck Josie" scrawled in black marker across the liquid crystal display (LCD). He'd been warned to keep her on the radar screen - Randy said she was a fast-tracker who'd end up overtaking Oscar if he wasn't careful.

But the real reason he was currently stumbling over the small talk was that he'd recently discovered she was beautiful.

Before he dug himself into a hole, he cut to the chase.

"Well, the reason I am here is the MixItMaster. Another customer reported that funky error. It's a little different, but I think it's the same problem."

Josie straightened, instantly on alert. Oscar had been a little more laid back because the problem happened so rarely, but he knew this bug bothered Josie because she couldn't find it.

"What happened this time?"

"Same behavior, different display message. Last time the machine stopped and displayed 'al Error', but this time it displays 'lume Fault'." He knew a barrage of questions was coming and quickly added, "It's the 5-liter version of the MixItMaster this time, not the 1-liter, so maybe that's why the error is different."

She dipped her head in silence and he tried to read her body language as she disappeared into engineer mode.

After several moments, she asked, "Who's the customer? The one who complained before?"

"No. GetOutOfTheLab has been using the mixer for seven months and never had a problem." He risked adding, "They told me they've seen it twice, although it hasn't messed up any testing. They're not upset, but just reporting a strange event. Otherwise they really like the mixer."

"That's good," she commented distractedly. "What materials profile were they running?"

"Viscous."

"Well, that's the same as the other place. Or solids. I think that profile also caused the bug once."

He continued to look around her cube while she pondered, and soon enough she had more questions.

"What were they doing when it happened? Had they just put more material in the mixer?"

"Well, they'd started with an empty hopper and loaded resin and fluid, then started a viscous mix cycle and walked away. After about five minutes they came back and the display said 'lume Fault'."

Josie clasped her hands and closed her eyes to think.

He wanted to help her, and tried to remember anything else that might be relevant. The MixItMaster was a tabletop laboratory instrument that mixed various liquid and solid raw materials to the appropriate consistency while heating the mixture to a specified temperature. After the mixing cycle completed, the user could press a button to dispense a preset amount of the material though a nozzle. The instrument had a little control panel with a keypad and a small one-line, 16-character display.

He had a thought. "If they add chunks of resin, shouldn't they use the 'solids' mixing profile?"

"No, if it has any liquid in there at all, they're supposed to use 'viscous' because the mixture can get pretty thick."

He couldn't think of anything else. The customers loved the instrument and Hudson was selling the hell out of it. The only problem was this bizarre message that had been reported twice.

"Can I talk to the person who saw it at GetOutOfTheLab?"

"Sure." He stood up to leave and promised to email her the customer's contact info. "We're planning a new release of the user manual. It'd be great if we knew what caused this before we go to press."

Josie hung up the phone and pondered the two new pieces of information she'd learned from the customer. First, the technician at GetOutOfTheLab had loaded the hopper with a large block of resin cut into chunks and a liter of viscous material. It was the same material used the last time this bug was reported, and she had been unable to reproduce the problem under those conditions.

She also learned that the mixer had stopped its mixing cycle prematurely. Mike hadn't relayed this vital piece of information, but that's why she liked to interview the problem reporter and not just rely on the messenger. A quick look in the defect tracking system showed the mixer stopped prematurely the last time, too.

Some bizarre set of conditions was sending the software into a fit and she couldn't figure out the set of inputs that caused it. She'd even tried playing "bad user" by randomly hitting buttons during the mixing cycle, adding materials at the wrong time, and tipping the whole machine as it ran. She'd nearly worn an entire hopper full of warm goop down the front of her shirt for that last trick, but she righted it in time and the machine had continued to run beautifully.

Rocking in her desk chair, she groaned at the tenacity of this little bug that needled her just when she'd managed to convince herself it was a fluke.

But now it was a battle. She decided she going to find the little bastard.

As she stood to walk out of her cube for a soda, her eyes lit on Li Mei's List pinned to the wall. She stopped to stare at it. Li Mei had formatted her hard-earned guidelines into bulleted sections using a nice handwriting font, and had printed copies for the team. Unpinning it from the wall, she started to read as her feet carried her on autopilot to the vending machines.

The bug was most certainly sporadic. *Really* rare.

She itched to open the listings but stopped herself. It hadn't worked last time and, besides, Li Mei had Josie's "Resist the urge to jump blindly into the software listings" quote right in black and white.

She chuckled.

So. What did she know?

- The same software supported both the 1-liter and 5-liter instruments, using a hardware configuration jumper to differentiate the two. Much of the code was shared and the bug had been observed on both instruments.
- The 16-character LCD display had a messed-up message; only about half of the characters were displayed.
- Both display messages were truncated and shifted all the way to the left of the LCD, displaying the ends of two valid messages ("Material Error" and "Volume Fault").
- The bug had been reported during the viscous and solids mixing profiles, but not the liquid profile.
- When each customer saw the bad display message, the motor was not running.
- The mixing cycle had ended prematurely.

From these symptoms, what could she brainstorm about the root cause for this sporadic display bug?

Since the instrument stopped mixing before the cycle was complete, it must have gotten into an error condition just before or after the bad display message was presented. After all, both times the message reported an error condition. But she didn't know if the error condition caused the display bug or vice versa. She wondered what could make *any* message appear shifted as the customers had reported.

Maybe it was something wrong with the display driver chip. The bug was elusive enough that it could be a bad part or solder short across two segments or pins that controlled the write position. Was an error condition missed or unhandled somewhere? Or perhaps the display driver was being told to write the characters to the wrong position.

She should also consider differences in the mixing algorithms for solids and viscous materials as compared to the liquids. Could something about those mixing profiles be borderline enough to trip an error condition? Memory corruption was also a possibility, although the instrument had no real-time operating system or dynamic memory allocation. All the variables had fixed memory addresses in the user RAM section of the memory map.

Finally, she pushed back from her notepad. The list had given her a few ideas she hadn't considered the first time and she finally allowed herself entry into the code.

She summarized her ideas:

- Hardware - bad display driver chip, solder bridges.
- Instrument is in error condition before or after bug starts - what causes these errors?
- How could display be shifted regardless of error status or processing state?
- Unhandled error condition or boundary condition?
- What is different between mixing profiles that could induce error?
- Memory corruption.

```
(Reader Information: header and include information
removed)
```

```
1    /* Defines */
2    #define    FAULT              1
3    #define    YES                1
4    #define    NO                 0
5    #define    HIGH               1
6    #define    LOW                0
7    #define    ON                 1
8    #define    OFF                0
9    #define    VISCOUS_HEATED     1
10   #define    LIQUID_HEATED      2
11   #define    SOLIDS_HEATED      3
12   #define    SLOW               1
13   #define    MEDIUM             2
14   #define    FAST               3
15   #define    MIN_VOLUME         10
16
17   /* Offset to Display Message strings (below) */
18   #define    ATTENTION_REQUIRED_MSG      0
19   #define    MATERIAL_LOADING_ERROR_MSG  17
20   #define    THERMAL_ERROR_MSG           34
21   #define    SPEED_ERROR_MSG             51
22   #define    VISCOSITY_ERROR_MSG         68
23   #define    SPEED_WARNING_MSG           85
24   #define    VOLUME_FAULT_MSG            102
25   #define    MIXING_MSG                  119
26   #define    DISPENSE_COMPLETE_MSG       136
27   #define    PRODUCT_INFO1_MSG           153
28   #define    PRODUCT_INFO5_MSG           170
29   #define    PROMPT_START_MSG            187
30   #define    SELECT_TEST_MSG             204
31   #define    ADD_INGREDIENTS_MSG         221
32   #define    MIXING_COMPLETE_MSG         238
33
34   /* Banks 1 and 2 of messages                   */
35   const char text0[]  = " Attention Req'd";
36   const char text1[]  = " Material Error ";
37   const char text2[]  = " OVERHEAT ERROR ";
38   const char text3[]  = "   Speed Error  ";
39   const char text4[]  = " Viscosity Error";
40   const char text5[]  = "  Speed Warning ";
41   const char text6[]  = "  Volume Fault  ";
42   const char text7[]  = "    Mixing...   ";
43   const char text8[]  = "  Empty Vessel  ";
44   const char text9[]  = "  MixItLab  100 ";
45   const char text10[] = "  MixItLab  500 ";
46   const char text11[] = "   Press START  ";
47   const char text12[] = "Select Test Type";
48   const char text13[] = "  Add Materials ";
49   const char text14[] = "Mixing Complete!";
50
51   const unsigned char motor_profile_1Liter[7]=
52       {0,20,25,56,60,115,198};
53   const unsigned char motor_profile_5Liter[7]=
54       {0,15,25,45,65, 85,198};
55   const unsigned char viscosity_ranges1_Liter[3] =
56       {10,50,221};
57   const unsigned char viscosity_ranges5_Liter[3] =
58       {12,57,221};
```

```
(Reader Information: function redeclarations &
unrelated variables removed)
```

```
120  unsigned char temp_A2D;   /* temperature sensor */
121  unsigned char visc_A2D;   /* viscosity sensor   */
122  unsigned char speed_A2D;  /* speed reading      */
123  unsigned char level_A2D;  /* hopper level       */
124  unsigned char *motor_profile_ptr;
125  unsigned char *viscosity_ptr;
126  unsigned char temperature_f;
127  unsigned char motor_enable_f;
128  unsigned char msec_ctr;
129  unsigned char output_byte;
130  unsigned char tenth_sec;
131  unsigned char time_counter;
132  unsigned char flow_check_counter;
133  unsigned char foo;
134  unsigned char LCD_position;
135  unsigned char high_nibble;
136  unsigned char low_nibble;
137  unsigned char mixer_motor;
138  unsigned char error_f;
139  unsigned int  sequences;
140  unsigned char temp_spike;
141  unsigned char minutes_since_mixed;
142  unsigned char motor_speed;
143  unsigned char tcal001;
144  unsigned char test_to_run;
145  unsigned char i;
146  unsigned char tmp;
147  unsigned char volume;
148
149  /* Port Information
150  Port A: I/O - general and LCD input/output
151  Port B: A/D - analog to digital conversion inputs
152  Port C: I/O - LCD and button, control I/O
153    portc.0 - portc.3: LCD display interface
154    portc.4 - product type (I)
155    portc.5 - start button (I)
156    portc.6 - heater enable (O)
157    portc.7 - motor enable (O)
158  */
159
160  #define LCD_E    porta.6   /* LCD Enable line */
161  #define LCD_RS   porta.7   /* Register select */
```

Figure 8-1 MixItMaster Partial Software Listing.

Reader Instructions: Review the code listings in Figure 8-1. Software runs on a small 8-bit microcontroller with one interrupt and onboard analog-to-digital converters to read motor speed, hopper temperature and fill level, and material viscosity. No RTOS. The microcontroller drives an external 1-line 16-digit LCD display, a single motor and a heater. Inputs include a small keypad with buttons to select the mixing cycle and to dispense the material. Like much code you will encounter, this isn't great but it's functional.

```
162  /* 16 digit LCD display functions */
163
164  /* Pulse Display Driver Enable line to tell
165     LCD driver new information is available.*/
166  void LCD_E_pulse(void)
167  {
168      LCD_E = LOW;
169      LCD_E = HIGH;
170  }
171
172  /* Tell LCD display driver where to start
173     printing on the 16 character display.
174     Position command sent high then low nibble */
175
176  void LCD_set_position()
177  {
178      LCD_RS = LOW;  /* Info coming is a command */
179      foo = LCD_position;
180      foo = foo >> 4;
181      portc = portc & 0xF0;
182      portc = portc | foo;
183      LCD_E_pulse();  /* pulse hi nibble over */
184      foo = LCD_position & 0x0F;
185      portc = portc & 0xF0;
186      portc = portc | foo;
187      LCD_E_pulse();  /* pulse lo nibble over */
188      LCD_RS = HIGH;  /* End sending a command */
189  }
190
191  /* This routine prints 1 character on display
192     and jumps the memory gap if needed. */
193  void LCD_print_char()
194  {
195      /* Advance RAM address at center of display.
196         Address goes: 80...,87,C0,...C7 */
197      if(LCD_position > 0x87 && LCD_position < 0xC0)
198      {
199          LCD_position += 0x38;
200          LCD_set_position();
201      }
202
203      /* Separate high and low nibbles  */
204      high_nibble = output_byte;
205      high_nibble = high_nibble >> 4;
206      low_nibble = output_byte & 0x0F;
207
208      /* Send high nibble */
209      portc = portc & 0xF0;
210      portc = portc | high_nibble;
211      LCD_E_pulse();
212
213      /* Send low nibble */
214      portc = portc & 0xF0;
215      portc = portc | low_nibble;
216      LCD_E_pulse();
217
218  /* increment display position variable to
219     match auto increment in display
220     driver chip.*/
221      LCD_position++;
222  }
```

```
223  /* Print all 16 characters on LCD display */
224
225  void LCD_message_print(message_num)
226  unsigned char message_num;
227  {
228      clear_display();
229      LCD_position = 0x80;  /* start at screen left*/
230      LCD_set_position();
231      interrupt_disable();
232      {
233          ;
234  #asm
235          LDX    message_num  ; get start of message
236  M1      LDA    text0,X      ; get next character
237          TSTA                ; check for null
238          BEQ    M2           ; exit if mssg end
239          STA    output_byte  ; store character
240          STX    tmp          ; save x register
241          JSR    LCD_print_char ; print character
242          LDX    tmp          ; restore x register
243          INX                 ; point next char
244          BRA    M1           ; display next char
245  M2      NOP
246  #endasm
247      }
248      interrupt_enable();
249  }
250
251  /* Read and adjust temperature/speed if needed */
252  void thermal_update(void)
253  {
254      read_hopper_temp();
255      if (temp_A2D > 136)
256      {
257          if (motor_speed > 5)
258              motor_speed-=5;
259          if (temp_spike++ > 10)
260              evaluate_heat();
261      }
262      compute_new_heater_temp();
263  }
264
265  /* Error occurred.  Turn motor/header off.
266     Display error message. */
267
268  void err(unsigned char message)
269  {
270      temperature_f = NO;
271      motor_enable_f = OFF;
272      heater_control();
273      motor_control();
274
275      LCD_message_print(message);
276      time_counter = 0;
277      while (time_counter < 120)
278          ;
279      LCD_message_print(ATTENTION_REQUIRED_MSG);
280      while (1)
281          ;
282  }
283
```

Figure 8-1 MixItMaster Partial Software Listing (continued).

Take a moment now and grab a pencil - don't be afraid to mark up the code! Choose a method, and figure out the code (flowchart, function calling trees, block diagram, etc.). You'll use your notes for the remainder of the mystery. Identify good and bad aspects of this code. (Some functions are not shown; consider the software they contain irrelevant to this mystery.)

Josie will shortly find an "aha!" in the display code related to her brainstorming list - can you find it first?

```
284  /* Mixing occurs here.  Motor cycles ON/OFF. */     347  /* A hardware jumper (Port C) defines this as a
285  void begin_mixture(void)                             348  one or a five liter machine.  Read jumper and
286  {                                                     349  then run appropriate function below (1L vs 5L) */
287      do {                                              350
288          motor_enable_f = ON;                          351  void select_product_type(void)
289          time_counter = 0;                             352  {
290          while (time_counter < 20)                     353      tmp = portc & 0x10;
291              ;                                          354      if (tmp == 0x10)
292          motor_enable_f = OFF;                         355          MixItLab_1liter();
293          time_counter = 0;                             356      else
294          while (time_counter < 5)                      357          MixItLab_5liter();
295              ;                                          358  }
296          interrupt_disable();                          359
297          check_volume();                               360  void MixItLab_1liter(void)
298          if (volume < MIN_VOLUME)                      361  {
299              err(VOLUME_FAULT_MSG);                    362      LCD_message_print(PRODUCT_INFO1_MSG);
300          interrupt_enable();                           363      motor_profile_ptr = &motor_profile_1Liter[0];
301      }                                                 364      viscosity_ptr = &viscosity_ranges1_Liter[0];
302      while (sequences-- > 0);                          365      select_profile_1liter();
303  }                                                     366  }
304                                                        367
305  /* Initiate running the VISCOUS profile */           368  void MixItLab_5liter(void)
306  void run_viscous(void)                                369  {
307  {                                                     370      LCD_message_print(PRODUCT_INFO5_MSG);
308      LCD_message_print(ADD_INGREDIENTS_MSG);           371      motor_profile_ptr = &motor_profile_5Liter[0];
309      detect_material_loading();                        372      viscosity_ptr = &viscosity_ranges5_Liter[0];
310      LCD_message_print(PROMPT_START_MSG);              373      select_profile_5liter();
311      while ((portc & 0x20) != 0x20)                    374  }
312          ;                                             375
313      begin_mixture();                                  376  void main(void)
314  }                                                     377  {
315                                                        378      RSP();
316  /* Run user-selected profile on 5L machine */         379      system_init();
317  void select_profile_5liter(void)                      380      interrupt_enable();
318  {                                                     381      all_initialization();
319      LCD_message_print(SELECT_TEST_MSG);               382      tcal001 = 0;
320      test_to_run = debounce_selection();               383
321      LCD_message_print(MIXING_MSG);                    384      /* Run forever, switching between mix and
322      flow_check_counter = 0;                           385         dispense cycles */
323      minutes_since_mixed = 0;                          386      do {
324      switch (test_to_run)                              387          switch(tcal001)
325      {                                                 388          {
326          case VISCOUS_HEATED :                         389          case 0 : select_product_type();
327              temperature_f = YES;                      390              LCD_message_print(MIXING_COMPLETE_MSG);
328              sequences = 480;                          391              tcal001 = 1;
329              motor_speed = motor_profile_ptr[2];       392              break;
330              run_viscous();                            393          case 1:  allow_dispense();
331              break;                                    394              LCD_message_print(DISPENSE_COMPLETE_MSG);
332          case LIQUID_HEATED :                          395              tcal001 = 0;
333              temperature_f = YES;                      396              break;
334              sequences = 240;                          397          default:;
335              motor_speed = motor_profile_ptr[4];       398          }
336              run_liquid();                             399      }
337              break;                                    400      while (1);
338          case SOLIDS_HEATED :                          401  }
339              temperature_f = YES;                      402
340              sequences = 600;                          403  /* Make sure hopper material level is within
341              motor_speed = motor_profile_ptr[3];       404     range before starting test */
342              run_solids();                             405  void detect_material_loading(void)
343              break;                                    406  {
344          default:;                                     407      if (level_A2D < 50 || level_A2D > 210)
345      }                                                 408          err(MATERIAL_LOADING_ERROR_MSG);
346  }                                                     409  }
```

Figure 8-1 MixItMaster Partial Software Listing (continued).

She opened the once-familiar file and immediately wanted to clean it up and move functions around to make it easier to read. The joys of inherited code. Sadly, it was pretty damn stable in this condition, and would thus remain unchanged unless she found a good reason to "fix" it.

She scanned the display routines, looking for missing error checking and possible boundary violations, and quickly became mired in the calling structure. It was mixed C and assembly. Shortly, she found the LCD code that printed user messages on the display, and she reverse-engineered it again.

```
410  void read_hopper_temp(void)
411  {
412      if (temp_A2D < 136)
413          LCD_message_print(SPEED_WARNING_MSG);
414  }
415
416  /* Based on profile, throw error if too hot */
417
418  void evaluate_heat(void)
419  {
420      if (test_to_run == VISCOUS_HEATED ||
421          test_to_run == SOLIDS_HEATED)
422      {
423          if (motor_speed < motor_profile_ptr[1])
424              err(THERMAL_ERROR_MSG);
425      }
426      temp_spike = 0;
427  }
428
429  /* Turn heater on or off */
430  void heater_control(void)
431  {
432      if (temperature_f == NO)
433          portc = portc & 0xBF;
434      else
435          portc = portc | 0x40;
436  }
437
438  /* Turn motor on or off */
439  void motor_control(void)
440  {
441      if (motor_enable_f == OFF)
442          portc = portc & 0x7F;
443      else
444          portc = portc | 0x10;
445  }
446
447  /* If goop sits too long without mixing,
448  turn motors on a bit to give it a little mix. */
449
450  void settling_control(void)
451  {
452      if (flow_check_counter > 60)
453      {
454          minutes_since_mixed++;
455          if (minutes_since_mixed > 25)
456          {
457              motor_speed = motor_profile_ptr[3];
458              motor_enable_f = YES;
459              minutes_since_mixed = 0;
460          }
461          flow_check_counter = 0;
462      }
463  }
```

```
464  /* If mix viscosity is out of range, try new
465      motor speed; otherwise, throw error */
466
467  void viscosity_check(void)
468  {
469      read_viscosity_value();
470
471      if (visc_A2D < viscosity_ptr[0] ||
472          visc_A2D > viscosity_ptr[2])
473          err(VISCOSITY_ERROR_MSG);
474      if (test_to_run == VISCOUS_HEATED ||
475          test_to_run == SOLIDS_HEATED)
476      {
477          if (visc_A2D > 150)
478          {
479              if (speed_A2D >= 128) {
480                  motor_speed = motor_profile_ptr[2];
481                  LCD_message_print(SPEED_WARNING_MSG);
482              }
483              else if (speed_A2D >=225)  {
484                  motor_enable_f = OFF;
485                  LCD_message_print(SPEED_ERROR_MSG);
486              }
487          }
488      }
489  }
490
491  /* Periodic timer ISR - 1 msec intervals */
492  void TIMER()
493  {
494      reload_timer();  /* Reload for 1 msec */
495      msec_ctr++;
496      update_mix_motor();
497
498      if (temperature_f == YES)
499          thermal_update();
500
501      if (msec_ctr == 100)
502      {
503          msec_ctr = 0;
504          tenth_sec++;
505          read_ADCs();
506          viscosity_check();
507
508          if(tenth_sec == 10)
509          {
510              tenth_sec = 0;
511              settling_control();
512              flow_check_counter++;
513              time_counter++;
514          }
515      }
516  }
517
```

Figure 8-1 MixItMaster Partial Software Listing (continued).

The **LCD_message_print()** function was called each time a message was to be displayed. She'd seen references to it throughout the code. It received an unsigned char that was a #define for the start location of the message string in ROM. The list of #defines and message strings was near the top of the file.

She pulled out the LCD display driver documentation to refresh her memory on how characters were sent to the display (see Figure 8-2).

First, **LCD_message_print()** set the LCD position, telling the display chip to start printing on the left side of the display. Then the #asm segment controlled reading one character at a time from the message string. #asm also called the function **LCD_print_char()** to send each character to the chip until the entire message was displayed.

LCD DISPLAY DRIVER AND CONTROLLER

1	2	3	4	5	6	7	8		9	10	11	12	13	14	15	16	◄——— Character Locations on Display
00	01	02	03	04	05	06	07		40	41	42	43	44	45	46	47	◄——— Corresponding Display RAM Address

Pins:

RS	Register selection (0: instruction register, 1: data register)
R/W	(0: write, 1: read)
E	Enable line: Initiates data read/write
DB7 - DB4	Upper nibble of data RAM
DB3 - DB0	Lower nibble of data RAM

Note: For 4-bit operation mode, upper 4 bits sent to DB7-DB4, followed by lower 4 bits. Lower RAM nibble DB3-DB0 not used.

RS	R/W	DB7	DB6	DB5	DB4	DB3	DB2	DB1	DB0	Instruction
0	0	0	0	0	0	0	0	0	1	Clear display
0	0	0	0	0	0	0	0	1	-	Return home
0	0	0	0	0	0	0	1	I/D	S	Entry mode
0	0	0	0	0	0	1	D	C	B	Display on/off control (D:display, C:cursor, B:blink)
0	0	0	0	0	1	S/C	R/L	-	-	Cursor or display shift
0	0	0	0	1	DL	N	F	-	-	Function set (DL:data len, N: # lines, F: font)
0	0	0	1	............Address............						Set CGRAM address
0	0	1	............Address............							Set Display RAM address
1	0	............Data............								Write data
1	1	............Data............								Read data

* Execution time (max) (fosc = 270 kHz) = 37 us

Figure 8-2 Specifications for the LCD Display Driver Interface (based on [1]).

Looking at **LCD_print_char()**, she remembered that the microcontroller communicated with the LCD display driver chip using the lower 4 bits on Port C. Luckily, the driver had a 4-bit mode so the micro could split each character into two parts, sending the upper 4-bit nibble and then the lower 4-bit nibble. After each nibble, the code pulsed the Enable line using **LCD_E_pulse()** to tell the LCD driver that new information was available on the port, and the state of the RS line told the LCD driver if the nibbles were instructions or message data.

Under some conditions, the **LCD_print_char()** function sent the "Set Display RAM address" command to specify where to display the next character on the LCD. That function was called **LCD_set_position()**.

That function raised her suspicions.

She looked back at the documentation to understand the RAM locations used for the LCD display. Each character in the message was stored in a specific Display RAM location. The 16 characters weren't stored in contiguous memory - that was the funky gap she remembered. The first eight characters were written to 0x00 - 0x07, and the second to 0x40 - 0x47. That's why **LCD_set_position()** was called - to jump from 0x07 to 0x40. For contiguous display memory, the LCD driver automatically incremented the position after each byte was received, but when the micro wanted to write to a different location, it had to explicitly command a position change.

But why did the **LCD_position** jump the gap from 0x87 to 0xC0? She wondered if she had the right memory map, and then realized the upper bit wasn't data - it was the "Set Display RAM address" command. Setting that upper bit changed data byte 0x07 into command byte 0x07 (0x87) and data byte 0x40 into command byte 0x40 (0xC0). When the software detected that the driver had reached the memory gap, it added 0x38.

The code seemed to work correctly.

Aha! Li Mei's List to the rescue! It was confusing to check the logic with the memory RAM addresses and the 0x80 command added together, so she'd never noticed that the code that jumped the gap from location 0x07 to 0x40 had no boundary checking for **LCD_position** values *lower* than 0x07.

Grabbing her pencil, she blocked out the 16 character positions on her pad and carefully printed a valid and a buggy message below them (Figure 8-3).

Figure 8-3 LCD Display of Valid and Bad Messages.

It looked like the bad message was starting too far to the left by a few characters. She scanned the file for all write accesses to **LCD_position** but was disappointed to find only two instances. Both were inside the display code. For each, **LCD_position** was set to a valid value right before the "Set Display RAM address" command was sent to the display driver.

Then she remembered the memory overflow condition with the Friends-Finder Communicator and flipped back through the code to the declaration for **LCD_position**. Unfortunately, it was smack in the middle of a bunch of unsigned chars. Not likely to be overwritten by a rogue pointer.

With a sigh, she made a note to consider a Change Request against the **LCD_print_char()** function to perform boundary checking on the position value, but doubted it had anything to do with the problem.

Oscar was in a foul mood at the status meeting. It ended quickly as he stormed out without comment after receiving yet another text message. Probably Randy again. Ravi exchanged glances with Li Mei, who shrugged and stood to gather her belongings.

"What's up with that?" he asked, trying to provoke some speculation.

Josie looked up. "Like you never had a bad day with your boss?"

"Well, he doesn't have to take it out on us," Ravi told her.

"He didn't. You just think he did." She rounded back on him. "What is your problem with him anyway? You almost act happy when Randy gets into a snit with him."

"Randy doesn't like him. How would you know what it's like to have a manager who favors *some* people over others, when you're one of the *others*?" he retorted, feeling his anger rise. He hadn't meant to say it, but he knew Josie got special treatment.

She stiffened and her expression turned black as she carefully placed her lab notebook back down on the table.

She took a step closer to him. "Ravi, you *make* yourself one of the 'others.' You think he's never nailed me to the wall for stupid things I've done? You have no idea what happened last year." Her voice was rising. "If you don't trust me, fine. But do me a favor and stop acting like you're getting singled out, because what you *don't* realize is that Oscar shields us from Randy and most of the upper-management bullshit."

Ravi stood perfectly still, torn between saying something he'd regret even more and testing a tiny bit of what she was saying for merit. He wondered what had happened between her and Oscar. After several moments of mutual silence, he looked up to find her still staring at him, although her ire had faded somewhat.

"No one's perfect. Don't turn into another Benjamin."

With that, she turned and left him grasping for something glib to say to Li Mei.

Josie pulled into the Sushi House parking lot for lunch, wondering how the situation in the conference room had mushroomed so quickly out of control. It had been simmering below the surface for months but she'd ignored it, not really knowing why Ravi had such a hard time with Oscar. *With me, too,* she amended. Aside from occasional explosions she'd had with Oscar, after which they hashed things out and got back on track without a grudge, she thought she had a good relationship with everyone on the team.

But now she realized that Ravi didn't talk to her much outside of meetings and rarely came to her for technical advice.

Her emotions conflicted: part of her wanted him moved to a different team if he was going to make daily interactions strained, and the other wanted to reach out to him.

But the latter required two people; she couldn't do that alone.

Whatever . . . she pushed the confrontation aside and tried to refocus on the mixer display problem.

Finishing her soup, she thought about an idea she had during the status meeting. Since she had concrete information about some conditions when the bug occurred, she should document the calling structure of the software to understand when these errors happened. The listings sat on the chair beside her, and she retrieved them to begin a rough function call tree while she waited for Siew to bring her sushi.

> **Reader Instructions:** A calling tree is a visual aid showing which functions call which functions, like branches on a tree. Commercial products can create interactive function call trees automatically, making deciphering huge code easier. Josie will finish hers by hand before leaving the restaurant - create your own now. Capture all the functions in your calling tree before continuing.

By the time she was finished with lunch, she felt confident she had captured everything. It wasn't really *that* bad after she'd pulled out the display and error functions that were called everywhere. She ended up with four segments of code: **main()**, **TIMER()**, **err()**, and **LCD_message_print()**. (See Figure 8-4.)

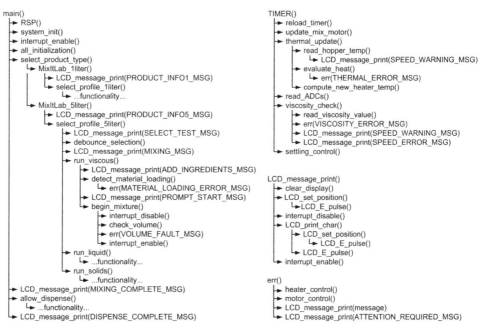

Figure 8-4 Calling Trees for the Software.

Since all three profiles were nearly identical except for the hard-coded stuff like text messages, motor speeds and mixing times, she decided to concentrate on the `MixitLab_5Liter()` code running the Viscous profile.

The bad message referred to a Volume Fault, and she found only one reference to that condition. The `begin_mixture()` code regularly checked the volume of the hopper, and if it fell below a minimum value, an error condition was called.

She peered back at the call tree. *Couldn't that only happen if someone physically removed material from the hopper during the mixing cycle?* She knew the mixing and dispense cycles were separate. The switch in `main()` toggled between the two, and the dispense valve would not open until the mixing cycle was complete.

It didn't make sense. The customer said they hadn't touched the mixer after starting the mix cycle.

On the other customer's bad machine, the truncated message was a Material Fault, which occurred when too little or too much material was added to the hopper. That condition prevented the machine from even *starting* the mix cycle. Throughout her pointed cross-examination, the customer *swore* it had mixed for a while before faulting.

Every symptom led to a dead end. She wanted to cry "Auugggh!" while laughing and groaning at the same time, but decided instead to leave the restaurant politely and head back to work.

She debated approaching Oscar but needed to bounce ideas off someone. Biting the bullet, she wandered over to his cube and tapped lightly on the edge of the opening.

"What." He commanded sharply without turning around.

"I was wondering if you are busy or could play soundboard with me." She tried to sound upbeat, but was prepared for the rejection.

Oscar turned and his face remained impassive.

"If you're busy, I can come back later." She started to back away until he waved her back.

"No, that's okay." He raked his hands through his short hair and exhaled forcefully, stretching as he stood up. "I need sustenance. Did you eat?"

"Yes, but that's okay." Trailing after him towards the cafeteria, she fished for change in her pocket to buy something to drink. Soon they faced each other at a far table overlooking the parking lot.

Josie avoided mentioning the status meeting and instead launched right into her dilemma about the mixer, first outlining the symptoms and then describing her subsequent brainstorming.

"And when I drill down each symptom, I end up looking at code that can't possibly be executing!"

Oscar continued to work at his sandwich. She stared out the window, mentally spent.

After a time he uttered, "Memory corruption."

She pulled out the linker symbol table information for the variables. (See Figure 8-5.) "I thought of that, but **LCD_position** is the variable that controls where the message is displayed. It's a global variable in the middle of the block with no pointers and no dynamic memory allocation - how could it get corrupted?"

```
SYMBOLS, continued - Version 3.69                PAGE 3

                    /* More variables, continued        */
    00CF            unsigned char temp_A2D;
    00D0            unsigned char visc_A2D;
    00D1            unsigned char speed_A2D;
    00D2            unsigned char level_A2D;
    00D3            unsigned char *motor_profile_ptr;
    00D4            unsigned char *viscosity_ptr;
    00D5            unsigned char temperature_f;
    00D6            unsigned char motor_enable_f;
    00D7            unsigned char msec_ctr;
    00D8            unsigned char output_byte;
    00D9            unsigned char tenth_sec;
    00DA            unsigned char time_counter;
    00DB            unsigned char flow_check_counter;
    00DC            unsigned char foo;
    00DD            unsigned char LCD_position;
    00DE            unsigned char high_nibble;
    00DF            unsigned char low_nibble;
    00E0            unsigned char mixer_motor;
    00E1            unsigned char error_f;
    00E2            unsigned int  sequences;
    00E4            unsigned char temp_spike;
    00E5            unsigned char minutes_since_mixed;
    00E6            unsigned char motor_speed;
    00E7            unsigned char tcal001;
    00E8            unsigned char test_to_run;
    00E9            unsigned char i;
    00EA            unsigned char tmp;
    00EB            unsigned char volume;
    00EC            unsigned char message_num;
    00ED            unsigned char message;
```

Figure 8-5 Symbol Table for the Final Portion of User Variables from the Linker File Memory Map.

"Dunno, but it does." He ignored the paper, working resolutely at the sandwich without looking up.

"Don't you want to look at the memory map or the code?" she asked.

"No. According to you, the instrument is behaving illogically, making memory corruption a logical conclusion. Are you blowing the stack?"

Now that was something she'd *not* thought of.

"I don't know."

"Don't forget that the messages were error messages and the instrument stopped mixing when the bug manifested. Check the max stack depth under those conditions."

She nodded. When it was clear he was finished talking, she quietly left him to the rest of his lunch.

> **Reader Instructions:** Use the function call trace in Figure 8-4 to identify the situation with the worst-case stack behavior - the deepest function nesting - before Josie finishes. Will this situation blow the stack? You'll need to identify stack usage each time a function is called. Use the information in Figure 8-6 and variable addresses in Figure 8-5.

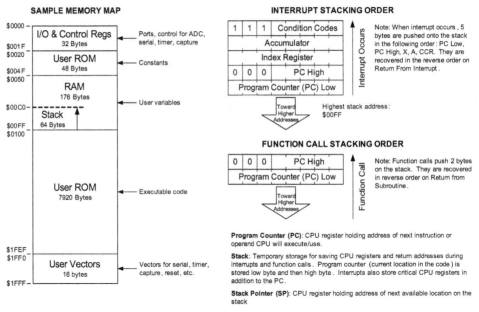

Figure 8-6 Sample Memory Map for the Microcontroller, based on the von Neumann Architecture. (Based on [2]).

Josie returned to her call trace, focusing on the worst possible nesting of functions during an error condition. If the instrument was in the middle of a mixing cycle and an error condition occurred, an error message was displayed. She counted 10 functions nested, ending with a final call to pulse the display's Enable line. (See Figure 8-7.)

Would that exceed the depth of the stack?

She dug around in her desk for the microcontroller's memory map. Unable to locate it quickly, she opted for the internet and typed the processor's part number into her favorite search engine. (See Figure 8-6.)

Function Call Tree: Deepest Nesting
```
main()
  ┣━► RSP()
  ┗━► select_product_type()
        ┗━► MixItLab_5liter()
              ┗━► select_profile_5liter()
                    ┗━► run_viscous()
                          ┗━► begin_mixture()
                                ┗━► err()
                                      ┗━► LCD_message_print()
                                            ┗━► LCD_print_char()
                                                  ┗━► LCD_set_position()
                                                        ┗━► LCD_E_pulse()
```

Figure 8-7 Josie's First Attempt to Identify the Deepest Possible Function Nesting.

From the microcontroller's manual, she found that the stack pointer was set to $00FF after power-on reset or a command to reset the stack pointer (**RSP**). After that, the stack pointer was decremented as items were pushed onto the stack. The RAM area was 176 bytes and shared the upper 64 bytes with the stack. That meant the stack could extend as low as $00C0.

She returned to the symbol table (Figure 8-5) and found that the last RAM variable declared in the software was located at $00ED. RAM had extended into the stack area, and she realized that could be a problem if the stack grew too large. **LCD_position** was also in the shared area at $00DD, although many other variables would have to be corrupted before it was.

Time to play computer.

Grabbing her pencil and her deepest function nesting, she counted out stack memory locations required for each of the 10 nested function calls. Each function call pushed two bytes onto the stack. Quickly, she found that the stack pointer would reach $00EB, which meant variables stored at RAM locations $00EC and $00ED would be overwritten.

Oscar was right!

On the other hand, **LCD_position** still remained unscathed.

She didn't let that deter her. Instead, she wondered if this stack problem could have caused any of the illogical symptoms. A quick search showed her that the overwritten variables **message_num** and **message** both stored indices to the text message to be displayed. She realized that meant completely random messages could be displayed. And both failures displayed error messages that she *knew* had been impossible for the conditions the customer described. One mystery solved!

Now she was completely immersed in the puzzle. It was looking less and less like a hardware problem, now that she had hold of the bug's slippery little tail.

Josie had unraveled many of the symptoms and was left wondering what had caused the error in the first place. It occurred only with viscous and solid mixing profiles. No, she amended, only with viscous and solid *inputs*.

Oscar's recent counseling returned to her. Identify the set of inputs that causes the unacceptable behavior, and then find what must be changed to make that behavior acceptable.

But the stack problem was hiding the true error message. Maybe she could tease it out by looking at code that generated errors. Where was **err()** called?

Function calling err()	What causes error condition?
begin_mixture()	- if the volume magically is too small during mixing
detect_material_loading()	- if too much or too little material is added before mixing
evaluate_heat()	- if the mixture becomes too hot during slow mixing for SOLIDS AND VISCOUS!
viscosity_check()	- if the viscosity is out of range during mixing

She immediately crossed off the first two items as unrelated, and concentrated on the last two possibilities. Looking at the list, she realized she should have asked the customers if they'd observed any heating or viscosity problems.

She returned to her calling tree and found that the last two errors were only detected during a TIMER interrupt. *Time for the big eraser,* she realized as she reconsidered her deepest nesting. *Get rid of the error message invoked by* **check_volume()**; *it's protected against interrupts. And add the TIMER ISR.* Belatedly, she remembered an equation she'd seen at the Embedded Systems Conference. For simpler systems like this one, the worst-case stack depth was equal to the depth of **main()** plus the worst depth of any interrupt. [3] She was embarrassed she'd forgotten all about including the TIMER interrupt.

Jumping back to the microcontroller's specs, she found that an interrupt pushed five bytes on the stack. Now her new deepest nesting drove the stack down to $00E2, blowing away several more variables, including motor speed. That could explain why the motor stopped if the drivers checked for bad values. But the stack pointer still hadn't quite reached **LCD_position**. (See Figure 8-8.)

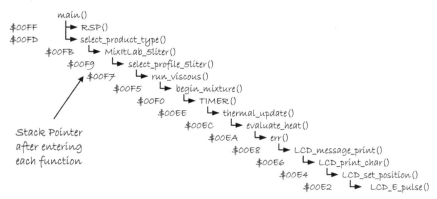

Deepest Nesting: Functions Call from main()

Figure 8-8 Deepest Function Nesting based on Viscous Mixing Cycle with a Thermal Error.

If the instrument detected a thermal error trying to mix viscous material, it would fault out and display a random message. She was pleased; one by one she was able to explain the symptoms with concrete information. She speculated that if the customer cut the resin chunks too large or didn't add enough fluid, air pockets could form and the air might overheat near the sensor and cause a thermal fault.

But, in spite of everything she'd discovered, she still had no idea why the display message was shifted too far to the left, and what a shifted display had to do with resin and thermal errors.

> **Reader Instructions:** Josie is on the right path: a thermal event occurred and the wrong messages are displayed due to stack corruption. The messages are being printed too far to the left, but she doggedly assumes that a corrupted **LCD_position** variable is the culprit.
>
> It's not. Can you figure out what is? You'll have to "play computer" to explore what happens to the stack during error conditions. This is your last hint.

She stretched, stiff in her seat, and heard Li Mei and Ravi talking in the hall. They wore jackets and Li Mei carried her ever-present red backpack.

Li Mei saw her turn and asked, "We're leaving now. Are you working late today?"

"Wow, I didn't realize it got so late." She pushed back and remembered she was meeting friends for dinner after work. Sometimes it was better to just let her brain work on the problem on its own while she relaxed, so she grabbed her jacket and followed them out of the building.

As she drove, nested function calls floated in her head. She tried to reorder them to make the stack grow longer, but they kept dancing away from her. Adding the TIMER, she'd nested as deeply as she could. Could it really nest any deeper to overwrite **LCD_position**?

Would overwriting any of those other variables cause the observed display shifts?

"This one is just freaking amazing!" Josie burst into the status meeting late, clutching a handful of paper and grinning broadly.

"I got the little bastard. Nailed him right to the wall!"

Everyone turned to look, and Oscar raised an eyebrow as if he'd been expecting her to solve the mystery *right about now*. Everyone in the aisle knew why she'd been holed up in her cube the last few days. They'd been treated to her one-sided adversarial conversation with the elusive bug as she proposed and rejected idea after idea. Li Mei had wanted to tape record her as a practical joke.

"Memory corruption? Stack?"

She nodded. "You were right, Oscar. Thanks for the tip."

"Well, since we're all pretty aware of the facts from your running hallway commentary," he drawled, seeing her cringe, "perhaps you'd like to share the specifics of your coup d'etat?"

She sighed dramatically and rolled her eyes at him. "Okay, but you're gonna love this one."

He noted the excitement in her gait as she walked to the whiteboard while outlining the original problem reports and symptoms. Her presence radiated confidence and her explanations were clear and easy to follow. She had solved this problem on her own with only a few pointers from him, and that made him feel proud and obsolete at the same time.

"So after I started dreaming about disabling and reenabling interrupts, I wondered if it were possible to have the TIMER interrupt itself and drive the stack pointer even deeper into the user RAM variables." She passed around a page showing how she believed the interrupt nesting had occurred. "A thermal event occurred and was detected within the first TIMER interrupt. It called the error code and the correct error message was displayed.

"Now if you look in the error code starting at line 268, you'll see what I missed the first 47 times. It prints the message correctly and then waits 120 seconds before printing a second message to get the user's attention. The #define for **ATTENTION_ REQUIRED_MSG** points to the message string "**Attention Req'd**." *That's* where the funkiness happens because it's still nested within the first interrupt."

She passed around the code for the **TIMER()**, **err()**, and the display functions (portions of Figure 8-1).

"As it prepared to print this second message, it calls **LCD_set_position** to start at the left side of the display. Then, the second TIMER interrupt occurs, nested in the first one."

Ravi jumped in. "I thought an interrupt service routine couldn't interrupt itself." He quickly relayed his interrupt overflow problem on the RoboGym module and explained that servicing the second interrupt was delayed until just after the too-long one finished.

"That depends on the type of interrupt and how it's configured. If you reenable interrupts inside the ISR and the ISR runs too long, the next one to happen will happily shove more stuff on the stack and begin executing."

"How can that happen if your interrupts are 60 seconds apart?"

"Oh, they're occurring every millisecond, but since nothing was being displayed, no bug manifested. But *this particular bug* required that one of the RAM variables be accessed somewhere else *after* its value was corrupted by the stack."

She faced them and paused to build anticipation.

"Is it **LCD_position**?" Oscar asked. He decided to play the straight man, since it obviously *wasn't*.

"Nope! It was **minutes_since_mixed**."

"What does that variable have to do with anything?"

"Ah, this is the cool part." Her eyes sparkled. "Timer interrupts occurred until **settling_control()** was called at line 450, which happens every second. Every sixtieth call to that function causes the variable **minutes_since_mixed** to be incremented, but *that* variable was *already overwritten* by the stack."

"How do you know what happened without looking at an execution dump of the stack during the error?" Li Mei asked her.

"I don't have a dump - this is all speculation, but I'm pretty sure it's right." Josie pointed to the pages she had passed around. "**minutes_since_mixed** is located at RAM address $00E5. If I work backwards . . ."

Li Mei interrupted loudly, "So *that's* what E5 is all about."

Josie looked at her askance.

"This morning you kept yelling, 'E5, it's E5!' " Li Mei snickered, and Oscar thought the corresponding sound clip would have been appropriate at this point.

"Yeah, I forget I talk out loud sometimes," she admitted. "But take a look at the final function-calling sequence for this bug and I'll explain it." She passed around the symbol table (Figure 8-5) and her final deepest listing (Figure 8-9), explaining that the left column contained the stack pointer value after each function was called. To the right of each function name, she listed the memory locations on the stack used to store the return address back into the calling function.

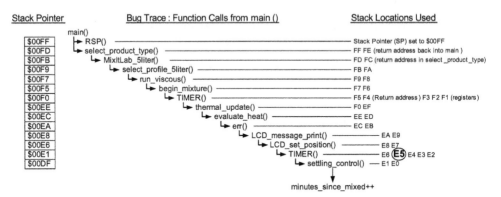

Figure 8-9 Actual Function Nesting Corresponding to the Shifted Display Bug.

"When each function is called, it pushes a return address on the stack. This lets the code know where to return to after it finishes executing the function. Each return address is two bytes long," Josie explained. She looked down at her own copy of the deepest nesting as she walked through the steps of the complicated chain of events. "Anyway, looking where $00E5 is overwritten by the stack, I see it was when the second TIMER interrupt occurred. $00E5 holds part of a return address that's from within the function **LCD_set_position()**. I figured out the program counter must have been right at line 179 where **LCD_position** is assigned to the temporary variable, **foo**, when TIMER #2 happened. And you can see that **settling_control()** is called in the second TIMER and it increments the **minutes_since_mixed** variable without error checking. HOWEVER," she emphasized dramatically, "**minutes_since_mixed** is *also* located at RAM location $00E5, so incrementing that variable actually incremented part of the address on the stack that was needed to return to **LCD_set_position** when the second TIMER code was finished."

Josie looked around the table and saw looks of partial understanding, and she decided to continue. "Finally, the second timer interrupt finishes and pops stuff off the stack. The program counter grabs the return address stored at E6 E5, but the contents have changed. This causes the program counter to jump *beyond* the line where **LCD_position** is stored in a temporary variable, **foo**."

Oscar shifted in his seat, wondering if he saw a flaw in her logic. "Josie, this increments the return address one byte, but each instruction takes two bytes in

ROM. It would return and start executing instructions off-by-one and crash immediately without printing the second shifted message."

"It depends on which byte is stored first in memory. This is a big endian machine, which means the high-order byte of a number is always stored at the lower memory address. The low byte goes on the stack first and then the stack pointer is decremented to a lower address before the high byte is pushed." She showed him the stack's order of pushes for interrupts. "So, it's not the low byte of the return address that gets incremented, but the high byte. Bumping that jumps the program counter forward about three lines of code. Each of those logical shifts to the right is a two-byte instruction."

She spread her arms for the finale. "Since the position assignment to **foo** and part of the position-setting code is skipped, a garbage value in **foo** is logically OR'ed with whatever is already on Port C, and the message starts printing way off the left side of the screen.

"Q.E.D."

Oscar was impressed.

After she'd walked through the bug again more slowly, Li Mei and Ravi told her they finally understood the sequence of events that had caused the elusive bug. Li Mei smiled widely and even Ravi decided to look impressed.

But the look on Oscar's face shook her to her core.

He looked at her like an equal. A peer.

She felt dizzy and disoriented, and she continued to stand stupidly at the whiteboard.

His expression didn't last, and he began grilling her about proving her hypothesis, in typical Oscar fashion. For this she was grateful, and began to relay her ideas for the solution.

"Even though I could induce the thermal problem by mucking with the materials, I can't force when interrupts occur, so reproducing the problem exactly would be difficult," she admitted.

"However, I *have* proved on paper that user RAM is overwritten by the stack, so I've got to separate the two. First, I'll remove unused variables and combine temporary variables where I can. This will reduce the amount of RAM needed for variables.

"Second, I *finally* get to attack the code and flatten the calling structure. That'll reduce the amount of stack space needed. And if I can't get all the user variables out of the shared 64 bytes of RAM that the stack uses, I will show deepest nesting on paper."

Li Mei said, "I don't understand one thing. What happened to the input variables for the LCD function calls? Weren't they supposed to be stored on the stack, too?"

"I was confused, too," Josie admitted, "until I found that function input arguments aren't saved in registers or pushed to the stack for regular function calls - only return addresses are pushed. Instead, the input parameters are converted to global variables during the compile/link process. I figured that out when the variable symbol table showed two more entries than the code's global variable list. That's dangerous, because you don't realize that user RAM is larger than you think, especially if you're having stack problems."

Li Mei nodded her understanding, so Josie finished her explanation.

"After I flatten the code and remove variables, I'll do a more formal stack analysis and run a bunch of tests."

Oscar grunted in approval. "Permission granted."

Li Mei looked up from the documentation. "It looks like you must remove at least seven levels of nesting. The last variable is stored at $00ED and the maximum stack depth leaves the stack pointer at $00DF."

"Probably only five functions because the TIMER pushes five bytes rather than two," Josie told her. "But, I have to prevent nested TIMER interrupts because most of the functions called by TIMER are not reentrant. They use static variables, like **foo**, that will be overwritten by the nested interrupt. All the display code isn't reentrant, and neither are any of the functions that check A/D values."

"Good job, Josie," Oscar said. "Write up the analysis and add it to the CR, and I'll get it assigned back to you for development. And don't forget about what causes the thermal problem in the first place."

Oh yeah, she conceded. *The unexpected input.* Then she remembered that the problem supposedly happened within a few minutes of starting the mix cycle. It could be an air pocket that prevented good mixing and caused a thermal event.

Still standing in front of the group, her brain was already off dreaming up fixes for bad material inputs. They could add a grid to the top of the hopper to limit the size of added material pieces. Or instead of immediately triggering a fatal error when high temperature was detected, perhaps displaying a "PLEASE HAND STIR" message or alternating the motor speeds. Or running the motor in reverse, or adding a second level sensor to detect air pockets - but where would she mount that second sensor?

"Earth to Josie, come in, please."

Everyone was staring at her.

Oscar repeated, "I want you to work with Ravi to flatten the code and then show him how to do the stack analysis. Are you okay with that?"

His question took her by surprise, but she nodded at both of them. She wondered if Oscar knew about the argument in the conference room.

Ravi looked over Josie's documentation and function call flows. She'd explained things thoroughly and answered his questions without making him feel like an idiot. Then she'd asked him to take a shot at flattening the code while she played in the lab with the materials.

> **Reader Instructions:** There are myriad ways to rewrite this code, depending on the end goals. Final revisions could vary radically if the team was under a tight deadline and needed to minimize risk, if they were allowed to start from scratch, or if the new executable needed to be downloadable in the field. Assume that Oscar has indicated that they are only to flatten the code structure and variables to address the stack problem.
>
> Try flattening the code. What functions would you eliminate and why? Will this fix the problem?

Josie had told him to look for instances of short functions that did little more than call other functions. They might be able to inline, or copy the contents of the more deeply nested functions into the level above, without significantly changing the architecture. She also recommended making sure functions were well-balanced with parallel structures.

He saw several candidates and made a list using her calling trees as a starting point.

Function	Reason to eliminate
LCD_E_pulse	Contains 2 lines, called 4 times. Limited to display code.
LCD_set_position	Called twice, only in display code.
LCD_print_char	Called only in display code (but from within assembly language segment).
read_hopper_temp()	Called only once - Parallel structure?
evaluate_heat()	Called only once - Parallel structure?
MixItLab_1liter, MixItLab_5liter	Calling function mostly empty.

Ravi summarized his results. It was clear that encapsulating functions for specific actions like **LCD_message_print()** shouldn't be unrolled into their calling functions. On the other hand, little things like enabling the display driver's E line and perhaps setting the position should be considered. He wanted to inline **LCD_print_char**, but that seemed risky since it was called from assembly code.

The TIMER code was nested three function calls deep. If he could unroll **read_hopper_temp()** and **evaluate_heat()** into **thermal_update()**, he could eliminate one level so that it had a structure parallel to **viscosity_ check()**.

Finally, looking at **main()**, he identified several levels of calls that just controlled which instrument code would run. After an hour of work, he was rewarded with a clean compile and link after eliminating seven functions:

```
MixItLab_1liter()
MixItLab_5liter()
select_product_type()
read_hopper_temp()
evaluate_heat()
LCD_set_position()
LCD_E_pulse()
```

He had unrolled the deepest nesting to seven functions calls and one TIMER interrupt. Now, the lowest the stack pointer would go was to $00EC, and a quick search showed he could eliminate the unused variable **mixer_motor**. Even if Josie couldn't remove any other variables, he had squeaked the stack just small enough that it no longer overwrote RAM. His only remaining concern was how to prevent the TIMER interrupt from reentering itself.

Whew! Now it was time to get Josie, and he found her at a shared testing bench staring at a mixing cycle in progress.

"I've finished," he told her. "Is the mixing working?"

When she saw him, she tossed up her hands and exclaimed, "When the chunks get large and I add just a bit of fluid, I can cause an air pocket, but I can't get the air pocket to form near the temperature sensor. I give up - I just wanted to reproduce this once and for all!" She shoved away from the bench. "What've you got?"

He handed her a new function call tree and his final deepest nesting (see Figures 8-10 and 8-11). "I can reduce the stack size to 19 bytes. It was 32 bytes when you started." He wondered how she would respond.

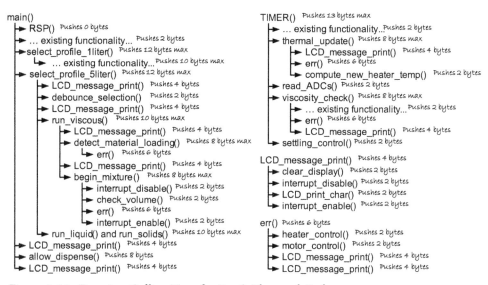

Figure 8-10 Function Calling Trees for Ravi's Flattened Code.

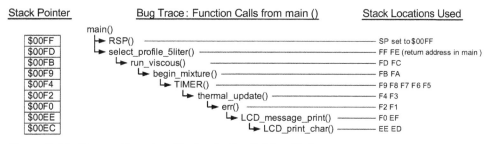

Figure 8-11 Deepest Function Nesting in Ravi's New Code.

"Thirteen bytes - that's a lot," she said doubtfully, looking at the sheet.

"Well, I don't know how to prevent the TIMER from reentering itself," he admitted, "but I am assuming that can be done. I eliminated seven functions."

"I'll take a look." She glanced back at the mixer, which had managed to whittle down the resin cubes and looked well on its way to a successful cycle. She flipped between his calling tree and the new deepest nesting pages. "OK, I see what you changed. Was it straightforward to select functions to eliminate?"

"Yes. What you suggested made sense - eliminating small functions and creating parallel formats." She didn't seem to be upset with him about the status meeting outburst, so he relaxed and added, "You know, it's pretty neat. After I unrolled several functions, I started seeing other strange lines of code that weren't obvious before."

"Like what?" she asked.

"Well, I unrolled **read_hopper_temp()** into its calling function **thermal_update()** (see Figure 8-12). The ADC value for temperature was checked with logicals in two different functions, but after inlining the code I can make the second one an else statement."

After dumping out the thick mixture, Josie wiped her hands on her jeans and accepted the code he held out to her.

```
322  void thermal_update(void)
323  {
324      if (temp_A2D < 136)
325          LCD_message_print(SPEED_WARNING_MSG);
326
327      if (temp_A2D > 136)
328      {
329          if (motor_speed > 5)
330              motor_speed -= 5;
331          if (temp_spike++ > 10)
332          {
333              if (test_to_run == VISCOUS_HEATED ||
334                  test_to_run == SOLIDS_HEATED)
335              {
336                  if (motor_speed < motor_profile_ptr[1])
337                      err(THERMAL_ERROR_MSG);
338              }
339              temp_spike = 0;
340          }
341      }
342      compute_new_heater_temp();
343  }
```

Figure 8-12 Segment of New Code Revealing Other Hidden Bug.

"You're right. Do that while I set up another mixing run," she said. "In fact, what if someone had changed one of the hard-coded 136s without realizing there was another one in a different function? Your change will fix that.

"Also, there's a bug. What happens if the temperature value is equal to 136? That condition isn't handled."

"I didn't see that," Ravi admitted. "I guess four eyes are better than two."

"Are four hands faster than two to get that mixer material chopped up?" Oscar asked, looking over their shoulders.

"Yeah, maybe you should grab a knife and get busy," Josie shot back at him, smiling.

"How's it going with the mixer changes?"

Ravi leaned against the bench. Oscar seemed in a better mood. "Good so far," he told him. "I just need to figure out how to prevent the TIMER from interrupting itself so the stack doesn't nest another five bytes. I reduced the nesting so RAM and stack don't overlap, but there's no buffer at all between them."

"That'll break the error function." Oscar stared at him.

Ravi stopped short, but told himself that Oscar was being matter-of-fact, not making a personal attack. So he steeled himself and casually asked why.

Oscar twirled his ID badge. "Because the error function still needs the timer to count time between the presentation of the error message and the second 'Come over here and see what's wrong with me' message."

"Oh, you're right," Josie groaned. "That means the stack can still blow into the variable RAM area. I want the system to be stack-safe - no chance of overflowing memory. But I haven't looked for useless variables to eliminate yet . . ."

"Why don't you just reset the stack pointer when you enter the error function? It's a terminal function, right? No need to return, no need to save a bunch of now-obsolete return addresses," he suggested. "Then throw a 'machine is in error' flag in TIMER to prevent it from running the motors and what-not, and it won't cause any more errors."

Ravi exchanged looks with Josie and they both nodded. It sounded like a good idea.

"Are there other ways to optimize the stack?" Ravi asked.

"On machines where function call arguments are pushed on the stack, reducing the number of arguments is a good idea. And pass pointers to arrays and structures rather than passing data itself," Josie told him.

"Even though inlining fixed this problem," she added, "inlining increases the size of the code. That could be a problem with this small memory footprint."

"No problem - we still have almost 2K left." He watched her cut up cubes of resin and feed them into the hopper through the prototype grate that Eduardo had created for her. "We could reduce TIMER reentrancy by speeding up any processing within the TIMER interrupt service routine. I had a problem with interrupts taking too long on the Animal project. Would that help here?"

She paused to think. "I don't know - you could profile the interrupt, but the reentrancy occurs in the error function when it's called from within TIMER. But you could try it. How did you do it before?"

Ravi felt a warmth in his stomach as he explained the pin-wiggling trick. A quick peek inside the closest MixItMaster revealed a socketed processor; tack-soldering a wire would be easy, and then she wouldn't have to configure the profiler.

"If you wait a minute, I can set it up." He saw her smile and found himself happy to show her something he had learned.

"Sounds great. Also, make a little function that fills the entire user and stack RAM space with 0x55 and have it run just after power-up but before the initializations. We'll inspect RAM after this runs a while and see how deep the stack nests."

"That will tell you how deep the stack can go?"

"No, it's just a best-case guess to understand stack behavior under normal operating conditions," she admitted. "The situation can get a lot worse with multiple interrupts running, and with recursion like our TIMER. Luckily, we're not running an RTOS like the Austin Monitor, because then each task has its own stack."

Ravi groaned. He noted her amusement and said, "That could get *so* out of control."

By the time Ravi had returned with an instrumented mixer, Josie had a pile of material ready. She was explaining to Eduardo why she thought using smaller cubes of material could address the problem.

"Maybe I could get a patent for my new grate - 'System Test solves software problem with mechanics'!" Eduardo posed dramatically when he noticed Ravi had returned.

She laughed. "I doubt it. And to be honest, the holes might need to be different sizes depending on the hardness or stickiness of the solids. I have no clue."

She accepted the instrumented mixer from Ravi and watched him set up the oscilloscope. As he worked, he explained the software changes to initialize memory and to toggle a spare output port pin for the scope.

Josie turned on the mixer and began adding materials until the hopper was nearly full, and then selected the viscous mix profile.

As they watched, the mixer began chunking the blocks of resin while the thick fluid flowed slowly around the blocks. Ravi fiddled with the scope dials to trigger on the start of each TIMER signal, and shortly they saw a rapid succession of pulses overlaid with their rising edges lined up neatly along the left side of the screen. He pointed to the rising edges. "This is the start of each TIMER interrupt, and the width corresponds to the execution time for the entire interrupt." After watching for several minutes he added, "The normal interrupt duty cycle is about 10 percent, and it jumps to about 15 percent several times a second."

Josie nodded, pleased, and explained to Eddie that the mixer had a really good duty cycle. It was unlikely to have reentrancy problems during normal operation because everything in the interrupt finished processing with a lot of time to spare before the next interrupt arrived. It was only during the terminal error code that the TIMER would interrupt itself.

She allowed the test to finish and was ready to check how deep the stack had gone, but suddenly realized they'd forgotten to connect the debugger.

"Ravi, we can't check the stack depth."

He broke into a grin and moved to the instrument's display panel, mysteriously pressing the START and DISPENSE buttons at the same time. She shook her head, wondering what he had up his sleeve. He was full of surprises today.

"I didn't want to lug the debugger over here, so I added a little stack monitoring code that runs when I press these two buttons. That code starts at the top of the stack and searches backwards until the stack value is *not* the magic 0x55 value and computes the stack depth." He pointed to the display. "I print the stack depth to the display, and also the previous five bytes so we can see what was last to be pushed on the stack."

"Hey, that's great!" Josie exclaimed. She wrote down the last two values, intending to look them up in the MAP file, a valuable debugging resource that listed the memory addresses for functions and variables. But it looked like the deepest nested function was in the initialization code.

"Ravi, this doesn't make sense." She showed him the file. "Deepest nesting can't be in initialization."

"That's the old memory map file. We ran my new code."

"That was the new flattened code? Running for the first time, on the hardest profile, without any code review, and it worked?" She broke into a grin. "Cool!"

"I'll bring up the new file." She followed him to his bench and looked over his shoulder as he used the deepest nested address location as a search string.

"Here it is," he said, pointing to the monitor. "The stack is only about twenty bytes deep and the last return address is in the `LCD_message_print` function where it calls `LCD_print_char`."

"That matches your deepest function nesting without the error call, Ravi. Nice job. Thanks for helping me fix this one." She handed back his function calling tree and saw Ravi's genuine smile in response to her praise.

She had been open and honest with him, and today it seemed to work. She found herself hoping that the positive interactions with him would continue.

Additions to Li Mei's List of Debugging Secrets

Specific Symptoms and Bugs

- Perform stack analysis during design to understand how much memory is needed. When stack size is limited, or when user variables are allowed to share RAM with the stack, stack overflow can cause catastrophic and unpredictable results.
- Suspect stack corruption when problems manifest in deeply nested functions or when a lot of data is passed between functions.

General Guidelines

- Play Computer. Doggedly step through the code line-by-line because sequential logic performed by a computer does not always match human assumptions. (Trace assembly language if necessary, noting the contents of registers including the stack pointer.)
- Use function call trees to understand the calling structure and function call nesting.
- Use debugging information generated by the tools (symbol tables, MAP and LST files, etc.).
- When reverse-engineering code, check off what functions you visit so you know where you've been. This can identify unused functions and variables.

Chapter Summary: The Case of the Impossible Symptoms
(Difficulty Level: Harder)

In this mystery, Josie is confronted by an elusive bug on the MixItMaster's LCD display: messages are shifted to the left. The symptoms all seem illogical until Oscar reminds her that illogical sporadic behavior can be caused by stack overflow and memory corruption. Using function calling trees, she verifies that the stack overwrites several user variables in RAM. The problem is fixed by flattening the calling structure and reducing the number of user variables so stack and user RAM no longer overlap.

The Problem Symptom(s):
- Truncated error message printed on display, starting off left side.
- Truncated messages were both valid error messages.
- Problem occurred on two different instruments using same software.
- Problem reported during mixing profiles for 'viscous' and 'solids' but not for 'liquid' materials.
- When truncated message was seen, the machine was not mixing but had not completed a mixing cycle.
- The problem was seen and reported by only two customers.

Targeted Search:
- Suspected problem with display driver (hardware solder bridge, bad chip).
- Questioned how the display message could EVER be shifted as observed, independent of current bug.
- Identified differences between viscous/solid versus liquid mixing profiles.
- Explored boundary conditions and error checking in display code.
- Memory corruption.
- Searched for causes of error conditions - is error a cause or an effect?

The Smoking Gun:
A quick check of the deepest chain of nested function calls showed the stack could overwrite the last few RAM locations.

The Bugs:
The software contained several problems, but the main issue was overlapping use of RAM.
- Theoretical stack depth during an error condition in a TIMER interrupt was $00E0; the last user variable was stored at $00ED. One of the last overwritten variables was a pointer to the error message, causing a totally unrelated message to be displayed.

- The truncated display message was caused when the LCD driver started printing at an invalid position because code to set the correct position had been skipped. This was caused when a variable, inappropriately overwritten by the stack pointer, was incremented during its normal operation. This increment corrupted a return address on the stack, causing program control to return to the wrong location in the code.
- An unrelated bug (mishandling of the value 136 in a logical) was found when the code was flattened.
- Proper processing of the error condition *required* nested ISRs. While this is not technically a bug, it is very questionable practice and should be avoided or documented heavily.

The Debugging Method Used:

- Interviewing the problem reporter.
- Creating function call trees to understand program control.
- Using symbol table and MAP files (generated by compiler/linker) to understand stack behavior.
- Brainstorming with a guru for a sounding board.
- Resisting the urge to blindly dig through software listings.

The Fix:

The final fix required three software changes and coordination between the Embedded and System Test groups.

- Flattened the calling structure so less stack space is needed to save return addresses for the deepest possible nesting.
- Removed several unused variables, reducing amount of user RAM needed.
- Identified new recommendations for addition of material to the mixer to reduce air pockets formation.
- Modified the mixing routine to avoid thermal errors.

Verifying the Fix:

- Stack behavior under "best case" conditions was investigated by using patterned memory, 0x55. (Verify using the symbol table that the first instances of 0x55 are not legitimate registers or return addresses.) True stack behavior lies between this best case value and the worst case (defined as deepest main nesting + the sum of all deepest interrupt nestings).

Lessons Learned:

- Detecting and compensating for all possible user inputs is extremely difficult.
- When a device behaves illogically, memory corruption is logical.
- With enough effort and insight (and a healthy dose of luck), bizarre symptoms can be unwound to an identifiable sequence of events.

Code Review:

Nesting is a large problem which is compounded by nesting of TIMER interrupts. Controlling what code can be performed in different states (mixing, dispense, error) isn't well handled (e.g., motor and heater are shut off in the error code, but TIMER still checks viscosity and material settling and modifies motor variables as if the device were running normally). Key debouncing is not always performed. Timing of message displays allows some messages to be over-written too quickly. Straightforward code is challenging to understand due to the hard-coded motor speeds, temperature values, and mixing parameters (use #defines and comments!). Hardware port pins are declared and handled inconsistently. While the implementation of the display messages may seem kludgy, this is common in ROM devices without operating systems.

What Caused the Real-World Bug? In the slower days of embedded systems, an embedded-systems engineer programmed an escape message into a device running a Gauss-Seidel iteration loop in case the loop proceeded longer than 20 minutes without finding a solution. The device displayed "HELP" to let the user know that no solution was possible. Years passed, and the code was made obsolete by a new algorithm that converged quickly, every time, and memories of the earlier cries of "HELP HELP" were soon forgotten.

The not-so-supernatural cause? The 8008-based instrument had a hardware stack seven entries deep. A long-buried bug finally surfaced that caused one PUSH too many, blowing the stack.

The moral - stacks are inherently problematic . . . document error messages! [4]

References

[1] Hitachi, HD44780U (LCD-II) Dot Matrix Liquid Crystal Display Controller/Driver Datasheet, *www.datasheet4u.com/html/H/D/4/ HD44780UA00FS_Hitachi.pdf.html*.

[2] Freescale, HC05 Family - Understanding Small Microcontrollers, *http://www.freescale.com/files/microcontrollers/doc/ref_manual/M68HC05TB.pdf*.

[3] Regehr, J. (2004), "Say no to stack overflow," *Embedded Systems Design*. *http://www.embedded.com/showArticle.jhtml?articleID=47101892*.

[4] Ganssle, J., Personal communication, August 1, 2006.

Additional Reading

Ganssle, J., (2000), *The Art of Designing Embedded Systems*, Newnes.

Real-World Bug [Location: Baltimore, Maryland, USA] A well-known embedded systems guru prepared to perform his annual boat maintenance. He filled his online shopping cart with antifouling paint for the bottom of his sailboat and several fuel filters.

His online order for maybe a hundred dollars worth of supplies appears below, totaling $83,929,999,999,999.98. Before tax.

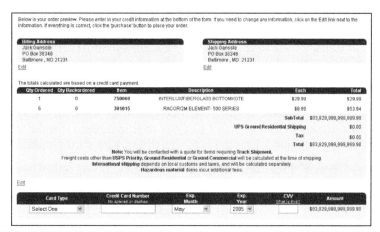

Chapter 9

When It's Not Hardware, It's Software. And Vice Versa. Blurring the Interface.

"So, Josie, to fix that mixer display bug, did you *really* have to prove that the truncated and offset display message was caused by a slightly corrupted return address popped off the stack after the high byte was accidentally incremented by the underlying overwritten user variable, thereby causing the display position to be programmed to start a bogus RAM address?"

She choked and hooted, holding her hand over her face to stop coffee from coming out her nose. "How long did you study the CR and practice that sentence before you could get it out in one breath?" After carefully setting aside the extra-large dark roast and wiping the tears from her eyes, she caught her breath and challenged, "And do you understand what any of that even means?"

Mike shrugged nonchalantly. Actually, it had taken him nearly an hour to rattle it off with any reasonable level of credibility.

"Nah. But Ravi tried to explain it yesterday afternoon. He said he would've just fixed the problem, but that you got on this rabid tangent trying to prove how individual little pieces of the program all lined up with the stars to cause that funky display message."

She cast an eye at Ravi and caught him eavesdropping. Although he tried to shift quickly back to the bench, she caught the smile on his face as he turned away.

What is it with these guys?

It was her turn to downplay. "Yeah, that bug was bothering me, " she admitted, "but I think it's fun to figure out exactly what caused a particular symptom. And that one was so strange that I just *had* to find it, even if it did keep me up a couple of nights. Sometimes I can get tunnel vision and a little obsessive about my work.

"In *fact*," Josie continued more resolutely, "I was on my way over to get obsessive all over Ravi about his test plan for these changes."

Hearing his name, Ravi finally acknowledged he'd been listening. "I know, Josie, I talked to Eduardo and he's going to help us with the testing. He's pulling more mixers out of storage so we can run tests in parallel."

"Good," she said. "With this last little bug fixed, I think we've got a winner."

"Terrific!" Mike smiled broadly at her. "I'm holding off the presses on the new user manuals. I'll get of the way so the geniuses can work."

———————————

"Hey, Ravi," Eddie taunted. "Gonna write a CR. This one's gonna be called, 'There's a little bit of crap right in the middle of the mixer display' and I'm making it a Severity 1."

Eduardo raised his eyebrows, anticipating an explosion, and was not disappointed. Hiding a grin, Ravi took off and grabbed an abused soccer ball resting on top of his filing cabinet, heaving it through the door. Eduardo flattened himself against the wall just in time, as the ball bounced down the hall and out of sight. Ravi tore careening out of the lab.

"I'll pound the crap out of you!" Ravi skidded to a halt, this time in much better spirits than the last time Eduardo had taunted him about code bugs. "You didn't find anything," he asserted.

Eduardo's expression sobered slightly. "Actually, we did, but it's the stupidest little thing. Come here - you gotta see it."

Eduardo walked back to the System Test lab with Ravi in tow, only to find the lab tech, Audrey, pouring a hopper fluid into the lab sink.

"Hey, Aud, what are you doing?" Eduardo was suddenly concerned. "I'm using that one."

"Just repeating Test 4.5," Audrey told him. "The test plan says it has to be done ten times."

Eduardo dropped to his knees dramatically and threw his arms in the air. "But that mixer had the display bug! I had to show it to Ravi!" Now he was in trouble - Ravi would think he made the whole thing up.

He looked up and pleaded, only half-mockingly. "I swear it's true. Right in the middle of the display was one letter that looked like some Chinese character or something."

"Sure it was." Ravi waved him off and turned to leave. "Nice try, Eddie."

"It was right in the middle of a normal message," Eduardo insisted as Ravi disappeared into the hall. He exhaled deeply, nodding distractedly as Audrey apologized for restarting the mixer. Pulling up the defect-tracking system with a sigh, he submitted the CR anyway.

Ravi plodded into the lab with the new CR in his hand and walked it over to Josie.

"Eddie found a bug in our code." He handed her the sheet. "It's assigned to me with your help."

"You're kidding." She sighed deeply and drained her coffee cup while she read. He felt as tired as she looked. He slid a stool over and then repeated everything Eduardo had told him about the "accidentally deleted" bug. He'd thought Eddie was pulling his leg.

"Which message had the bug?" she asked him. Unfortunately, Eduardo, in his excitement to implicate Ravi in a new bug, had forgotten to note the display message before Audrey reset the machine.

"Well, tell them to keep testing," Josie advised, "but to pay close attention to every screen that's displayed. That oughta keep Eddie out of trouble for a while. In the meantime, we should review the stack changes to see if the inlining is correct."

Shortly, they had the old code and the new code side-by-side on the table. Ravi leaned over the table with a finger on each, walking through lines of code to compare them. "I don't see any differences between them, functionally. All I did was copy the contents of the nested function calls directly into the calling function. Then I deleted the functions we decided to eliminate."

Josie tucked a strand of hair behind one ear and agreed. "Even the assembly language parts are the same." She tried to be encouraging, but neither had much energy after the late-night validation session they'd pulled so they could give Mike the approval to go to press with the new user manuals. "Let's do another stack analysis. As Oscar said, 'Illogical behavior makes memory corruption logical.'"

"I guess so. I'll get a debugger so we can see the whole stack. This one is going to be a pain to reproduce, too." Ravi reluctantly trudged back to the embedded lab.

Oscar walked purposely through the halls up Ritchie Way to Feynman, past his office, and rounded the corner to Kathy's cube. He was in luck - she was there.

"Kathy, I need to talk to you."

She greeted him warmly and waved him in. She was much more than the typical administrative assistant; she did everything for the team and always wore a smile. He had difficulty mustering that kind of energy on a regular basis and he admired her for it. Even if she *did* have pictures of cats hanging nearly everywhere in her cube.

He'd tormented her that *he* preferred his cats breaded and fried. She'd given him a deep-fried hairball for his birthday. Their relationship had been sealed.

But today she looked at him expectantly, and he realized he'd let his mind wander while he debated just how much he could reveal.

"I've got some news. First, you're being promoted with a nice pay raise. Second, our team will be expanding as we take on a big project." As Oscar waited for her reaction, he suddenly realized he really wanted her buy-in. "We're going to need you. I need you," he added.

"Really? Oscar, that's a surprise. Thank you!" Her face broke into an even wider smile. "Congratulations on the big new project. Is it what you want, I mean?"

"Yes," he assured her. "Hudson Technologies will be having a reorganization and some things are going to change. The bigwigs will announce things at a Town Hall meeting for everyone on Friday, but I wanted to let the team know what to expect beforehand."

Her face clouded slightly. "Is anything bad going to happen?"

"Well, a few folks will be moved around, and we're branching into some new areas. But I haven't heard anything about layoffs."

"That's good to hear. Is this what the late nights have been about lately?"

"Some of it." He tried to straighten his legs but he risked hitting her satchel so he crossed them instead. "Not everything is public yet, but we're not reporting to

Randy any longer." He saw her raise an eyebrow but she said nothing. She wasn't stupid. He knew she'd seen Randy downplay some of the team's successes.

"At least Randy fought for the team to go to that Embedded Systems Conference a while back," she offered.

"No." Oscar grunted his annoyance. "It was Mike who pushed hard after our success at the Telecom trade show. He's got the ear of the financial guys, who did some sales projections for the Austin Monitor after that. They figured it was a no-brainer to reward the team - an 'investment in our future' is what they called it. So Randy was forced to go along with it." He added, "Mike's being promoted as well."

"Well, isn't that something. I'm delighted!"

He swore Kathy to secrecy until the announcements were made. It felt good to give her the news and see her excitement. Nervousness over the coming announcement and changes kept him from sharing her wholehearted enthusiasm.

Ravi opened the defect tracking system. The CR was in the Analysis state, and he began typing their progress. They had reviewed the code changes again and found nothing, and a day in System Test showed no memory corruption or stack overflows. Or any further incidence of the bug. He finished by requesting that the CR be reduced to a Severity 3.

Later that afternoon, Ravi found Josie in the System Test lab, her head down and resting on her arms. She was taking her turn babysitting the mixers. As he approached, she still didn't move, and he wondered if she had actually fallen asleep.

In front of her, the mixer still churned, a dirty yellow conglomeration of thick goo and chunks that pulsed in the hopper. He looked at the display to check the status, and what he saw surprised him. (See Figure 9-1.)

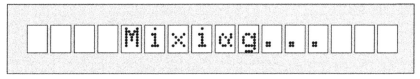

Figure 9-1 LCD Display Showing Bad Character in Message.

Wonderment coursed through him; Eddie had been telling the truth. Well, he had realized Eddie wasn't lying when the CR showed up, but it was a nice confirmation to actually see the failure.

Still, Josie hadn't moved.

He got an idea. Leaning in close to her ear, he took a deep breath and yelled, "Quick! Before it disappears! I can't believe you didn't see the bug!"

He jumped back as Josie leapt from the stool, which clattered on the floor behind her. Stunned, she looked abruptly between him and the mixer.

She *had* been asleep!

"What's your problem!?!"

"There's the bug," he said, trying to distract her. "Eddie was right."

"What?" She did a double take on the mixer's display and stopped in mid-response. "Wow - there it is. And it's just one garbage character like he said. Huh."

She stared for a moment and then righted the stool to sit back down. Still looking at the mixer, she suddenly rolled her eyes, "It's not garbage or a Chinese symbol, it's a real character. It's a Greek alpha. I don't even know how to print Greek characters on the display."

"How did an alpha get right in the middle of a normal display message?" he asked innocently. The bug had already reeled her back in.

"If it were a completely random bit pattern, it could mean hardware or bad soldering, but this is a legitimate character."

Li Mei called her greetings from the doorway and they looked up.

"I heard Ravi yelling. Is everything okay?"

Josie shot him a look.

"We're okay. But Ravi will need to watch his back for a while."

Ravi had the sense to look abashed, but was secretly pleased he'd been able to pull her chain. He quickly pointed to the mixer. "Come see the display bug Eddie found. It happened again on this mixer."

He described the testing and explained the difference between this bug and the last one. "Before, the message started printing too far off the left side of the display. This time, it prints in the correct location, but there's a random character in the middle."

Li Mei looked in. "That is interesting. An 'n' has been replaced by an 'α'. Right in the middle."

"Yeah, and this can't be memory corruption because all those display messages are hard-coded in ROM."

"Well, it could be corrupted after the code reads each character but before it displays the character," Josie warned. "That gives us somewhere to start. Let's go look at the code again."

Before they left, Ravi taped a sign to the mixer. "Do not touch! Do not unplug! Penalty is Death! Signed, Ravi the Enforcer."

After Josie had insisted on breaking for more caffeine, all three sat huddled around her monitor. They found the code that loaded individual characters from ROM and prepared them to be sent to the LCD display driver. (See Figure 9-2.)

"So the **LCD_message_print()** function sends each byte in the message, and the only change is that you inlined the **LCD_set_position** code, right?" Li Mei asked.

Josie and Ravi both nodded their assent. "It's the same. And it has to be right; otherwise, none of the displayed messages would come out correctly."

```
193  /* Print all 16 characters on LCD display */
194
195  void LCD_message_print(message_num)
196  unsigned char message_num;
197  {
198      clear_display();
199      LCD_position = 0x80;  /* start at screen left*/
200
201      /* START inlining old LCD_set_position() code
202         and old LCD_E_pulse() code */
203
204      LCD_RS = LOW; /* Info coming is a command */
205
206      foo = LCD_position;
207      foo = foo >> 4;
208      portc = portc & 0xF0;
209      portc = portc | foo;
210      LCD_E = LOW;      /* pulse hi nibble over */
211      LCD_E = HIGH;
212
213      foo = LCD_position & 0x0F;
214      portc = portc & 0xF0;
215      portc = portc | foo;
216      LCD_E = LOW;
217      LCD_E = HIGH;  /* pulse lo nibble over */
218
219      LCD_RS = HIGH; /* End sending a command */
220
221      /* END inlining old code */
222
223      interrupt_disable();
224      {
225          ;
226  #asm
227          LDX     message_num  ; get start of message
228  M1      LDA     text0,X      ; get next character
229          TSTA                 ; check for null
230          BEQ     M2           ; exit if mssg end
231          STA     output_byte  ; store character
232          STX     tmp          ; save x register
233          JSR     LCD_print_char ; print character
234          LDX     tmp          ; restore x register
235          INX                  ; point next char
236          BRA     M1           ; display next char
237  M2      NOP
238  #endasm
239      }
240
241      interrupt_enable();
242  }
243

244  /* This routine prints 1 character on display
245     and jumps the memory gap if needed. */
246  void LCD_print_char()
247  {
248      /* Advance RAM address at center of display.
249         Address goes: 80...,87,C0,...C7 */
250      if(LCD_position > 0x87 && LCD_position < 0xC0)
251      {
252          LCD_position += 0x38;
253
254          /* old LCD_set_position start */
255          LCD_RS = LOW;
256          foo = LCD_position;
257          foo = foo >> 4;
258          portc = portc & 0xF0;
259          portc = portc | foo;
260          LCD_E = LOW;
261          LCD_E = HIGH;
262
263          foo = LCD_position & 0x0F;
264          portc = portc & 0xF0;
265          portc = portc | foo;
266          LCD_E = LOW;
267          LCD_E = HIGH;
268          LCD_RS = HIGH;
269          /* old LCD_set_position end */
270      }
271
272      /* Separate high and low nibbles  */
273      high_nibble = output_byte;
274      high_nibble = high_nibble >> 4;
275      low_nibble = output_byte & 0x0F;
276
277      /* Send high nibble */
278      portc = portc & 0xF0;
279      portc = portc | high_nibble;
280      LCD_E = LOW;
281      LCD_E = HIGH;
282
283      /* Send low nibble */
284      portc = portc & 0xF0;
285      portc = portc | low_nibble;
286      LCD_E = LOW;
287      LCD_E = HIGH;
288
289  /* increment display position variable to
290       match auto increment in display
291       driver chip.*/
292
293      LCD_position++;
294  }
```

Figure 9-2 Listing for Display Code.

Reader Instructions: Their code change caused the bug. You have all the information you need in this chapter and the previous chapter to hypothesize how it happened.

"It doesn't *have* to be right," Li Mei replied. "You changed the code, and that changed the stack nesting, for one thing."

"But we checked the stack and it's fine," Ravi told her. "Besides," he reminded them, "Eddie saw the bug with the liquid profile. That's different from the last bug."

"Something changed," Li Mei insisted. "And Josie just saw it using the viscous profile. So *what* are the causes of sporadic bugs like this one?"

Ravi was amazed. She was lecturing him? Well, he amended, she was just being Li Mei, so he sighed theatrically and drawled, "*Okay* . . . I'll get The List."

"I will tell you, since I have it memorized," she said, and began reciting. "Sporadic events are seemingly random failures that can be caused by a variety of things including boundary conditions, unexpected inputs and outputs, unhandled error conditions, faulty logic, hardware, timing, memory corruption, and performance issues.'"

Josie raised her eyebrows at the impromptu lecture, but nodded her support. "The last bug was memory corruption." She turned back to the monitor. "I also realized a possible boundary condition issue with the display RAM gap when I was checking Ravi's changes. Check out the **LCD_print_char()** function."

She handed them a page and Ravi saw it was the LCD display driver documentation. (Chapter 8, Figure 8-2).

"From the documentation, the first eight characters get stored to LCD RAM at address 0x00 through 0x07, and the second eight are stored at 0x40 through 0x47." She cautioned them, "Don't get confused by the 0x87 and 0xC0 in the software. The display command to set the new position has the highest bit set to '1'. To make it easy, we just add 0x80 to every position byte memory location we send."

"Ahh, that's why it starts at 0x80 and then jumps the gap to 0xC0. I get it now," Ravi exclaimed.

Josie continued, "Line 250 has no error checking on the position value. I originally thought that was our bug."

"It might still be, Josie." Li Mei got up. "Wait a second."

She returned a moment later, excited. "The bad character on the display is at the ninth location out of 16. That is the start of the second half of memory, at location 0x40. That's a boundary."

Josie grabbed a calculator. "After the position command is incremented from 0x87 to 0x88, next time through the routine the code adds 0x38 to jump to 0xC0. That bridges the memory gap appropriately. The garbage character is printed in the right position, so I'm not sure that's the problem."

Li Mei pouted, "But it's the boundary where the bug occurs."

"True . . ." Josie admitted. "Is anything done differently with the display character when we jump the gap?"

Ravi looked back at the code. "No. And the memory location for `output_byte` is pretty far back in the map. Probably not corruption."

They fell into silence again. Li Mei left the room, returning sometime later with a cup of tea and a bag of pretzels that she offered to the group. Taking her seat next to Ravi, she leaned over to him and whispered, "I was just in System Test and saw your sign. The mixer you taped it to has been rotated exactly 90 degrees on the bench."

Ravi said nothing, feeling his ears burn. He vowed when this was all over that Eddie was going to suffer some indignity. He dreamed about updating Eddie's name on the door with random garbage characters in the middle, wondering if he could "accidentally" spell something of questionable taste.

He pursed his lips and grabbed the LCD driver documentation to look at the section on display character codes. Interestingly, many of the codes were actually normal ASCII codes. He looked up the code for 'n'; it was 0x6E. The character that had blown it away was 'α', or 0xE0.

"`Mixing...`" had become "`Mixiαg...`".

While Josie showed Li Mei their stack-testing results, he continued to puzzle over the character codes looking for patterns. He wondered if it was coincidence that both codes contained the hex value 'E'.

The thought nagged at him.

They used the 4-bit interface mode to the display. Each command and data byte was sent four bits at a time, high nibble/low nibble - the software was riddled with references to four-bit nibbles. After each nibble was sent, the Enable line was pulsed to tell the display driver that data was available.

He scratched his back and wondered if the two nibbles somehow got mixed up, but then realized that the garbage character would have been 'ρ' instead of 'α'.

He wondered what would happen if the '6' were sent and then lost and then 'E' was interpreted as the first byte sent. That would cause the alpha character to be displayed. After all, the software cleared Port C right afterwards, and that would provide the necessary low nibble to print 0xE0 or 'α'.

Ravi looked over the character chart and thought the idea made sense. Taking a deep breath, he dumped his idea to the group and waited for Josie to tear it apart.

But for the second time today, the expected thrashing did not come. Instead, she and Li Mei took his idea seriously and they actually debated how that particular sequence of events could happen.

"Aha!" Li Mei suddenly cried. "This bug *is* on the List - it is bad timing!"

Josie was right behind her. "Yes, this could be a timing problem. It's sending the upper nibble and pulsing the Enable line, but the display driver is still busy performing the change position operation to jump over the memory gap. The display starts listening again when the lower nibble is available, which is the hex 'E', and the display interprets that as the high byte."

"But there isn't another enable pulse after that," Li Mei pointed out.

"True, but who knows when the display is listening at that point. When it's busy, I don't know if it eventually reads the value or ignores it. By the time it responds to the second E pulse, our code may have already cleared the port in preparation for the next character."

Ravi was pleased that he had identified the problem, but was simultaneously embarrassed that he didn't quite understand it himself!

He asked, "Why is this just happening now with this software? Why didn't we see this before?"

Josie turned back to the screen and pointed to the code to jump the memory gap. "This is where you inlined the call to **LCD_set_position**. It now executes fewer instructions because it doesn't have to push the return address on the stack and jump to the new function. The display driver chip requires a certain amount of time to process each command, and you must have removed just enough instructions so it wasn't ready to receive the first nibble of the new character."

"No way." Ravi was amazed. "The timing is that critical?"

"Technically, we are supposed to check the busy signal from the display and wait before sending new information, but we don't have hardware access to that line. We've been lucky up to now. In fact," she added, "inlining code is something you usually do to speed up sluggish performance, but this time it made the performance too fast!"

"Is there any way to prove it?"

"Measuring that will be difficult to do in real time - we could run the profiler but instrumenting the code will affect the timing because the values are so small. We could also do it by hand, adding up the time per instruction."

Ravi had an idea. "I can create a custom version of code that just writes messages over and over every second. If timing is really on the edge, we should make the bug happen more quickly."

"Let's try that."

"Ravi, this code is throwing bad characters all OVER the place!" Apprehension clouded Li Mei's face. "This is a lot worse. I am concerned we have a more serious problem."

"Really?" Ravi asked innocently. He went to her bench and saw the display cycling through regular messages every second. About every tenth screen, a garbage character appeared somewhere in the message. He noted with satisfaction that the bad characters were not limited to the memory location just after the gap in the middle of display memory. He'd tried to be clever about forcing the bug to happen more often, and it had worked.

In a big way.

Several more screens went by, and one actually contained three bad characters.

By now, Josie had joined the group and was hunched over the mixer as well.

"What did you put in this code?" she asked, looking at him suspiciously.

Ravi smiled at her. Although he hadn't been on this project from the beginning, he felt he'd learned more than everyone else about the display hardware and the driver. The bit-level and hardware interface issues were more fun than the pure software aspects of embedded device development. Seeing the garbage characters jumping all over the screen cemented his hypothesis - they were sending data to the display driver before it was ready.

Now it was time to let everyone else in on his discovery.

"I added a debugging function in **main()** that just cycles through the display messages every second, because we thought we'd see the bug quicker with more frequent display updates. But that still might take too long, so I decided to stress the system a little." His heart skipped a beat and he paused to take a breath.

"The regular code explicitly changes the display position before the first character and the ninth character. For all the others, the display driver increments its own position pointer. We don't have to do anything. So that means timing is not an issue for any other characters. Then I wondered if I could make the bug happen more often if I explicitly commanded a position change for every single character instead of just at the gap. That's what this version of code does."

He watched the look of understanding dawn on both their faces and then he knew for sure he was right. Josie's expression was especially rewarding, and he basked in it.

"That was a great idea, Ravi. I didn't even think of it," she told him, "and I was worried whether we could reproduce the bug."

"I also thought if we can induce the bug this often, we could use this same debug code to test a fix," he added.

She seemed to like his hypothesis, and she asked if he had any ideas to fix it.

He had one idea, but it was an idea that had gotten him in trouble before, so he was hesitant to offer it.

"The LCD Display Driver spec mentions that the hardware driver needs about 40 microseconds to perform the operation. If we add a short delay after each command to change position, I think that would fix it. However, we're already using the timer and I'm not sure how to measure 40 microseconds accurately." He told her about the problems he had with the Animal project and the machine-dependent loop counter delays that had broken the system when it was upgraded to faster hardware.

"We could use a little assembly-language delay, but that got me in trouble before." He spread his hands in an expression of puzzlement.

"I see what you mean." Josie looked between them and quickly made a decision. "Before we worry about the fix, let's verify your hypothesis. Create a little loop in assembly language and place it after each character write to the display."

She borrowed a calculator from Eduardo's bench and punched at the keys. "For this system, each instruction takes about 0.5 microseconds, so we need 80 instructions to make 40 microseconds."

She motioned and Li Mei passed her a notepad. "Use an ASM loop with three instructions like this. Just load a loop counter in the accumulator and decrement it until it's zero."

```
#asm
                    LDA put number of loops here
        DLAY_LOOP   DECA
                    BNE DLAY_LOOP
#endasm
```

She explained, "If we need to burn through 80 instruction cycles, the accumulator should be set to about 26, based on the number of machine cycles required for this loop."

Ravi interjected, "But that's probably overkill if it worked most of the times before. That's adding a lot of instructions."

"You're right. But it guarantees that we wait long enough since we can't use the busy line to poll the display driver. Since this adds 40 microseconds per character, it will be an extra 640 microseconds per screen. No one's gonna miss that time."

Ravi thought about it but wasn't sure he agreed.

From Josie's body language, it appeared she was also having second thoughts, and it wasn't long before she amended her previous conclusion. "Well, we probably should do a more thorough stack analysis because that time is added to the TIMER interrupt."

"But," Li Mei interrupted, "TIMER only calls the display code on error conditions, and then we don't care because we're resetting the stack pointer."

"That's not true. A couple of messages are warnings, but normal operation continues," Josie cautioned. "And I just realized that 640 microseconds is about two-thirds of a normal 1-millisecond interrupt cycle - we're risking interrupt over-run. What we're dancing around right now is that the entire mess of code needs to be rewritten from scratch."

"But we don't have time for that," Ravi exclaimed. "Mike has that new release deadline."

"I know. But sometimes when software is not well architected and fixing one bug triggers other tricky bugs, you have to wonder what other time bombs are right below the surface. Maybe we just introduced a third bug while fixing the second one." Josie looked between them and asked, "Are you both confident that the entire code base running on this hardware is stable?"

Neither spoke, and Ravi realized she was right. He'd like to move more processing out of TIMER by having the main loop schedule actions based on time recorded by the interrupt. Maybe add message print management to `main()` and have the ISR set a flag with the message ID. That would also make the ISR less complicated to analyze and debug. He offered the idea and Josie suggested he create a new CR for analysis. "I also think we could just add the delay loop when we jump the gap rather than for every character - that's where the problem occurred anyway - then we only add 80 microseconds instead of 640 microseconds."

"That's smart thinking, Ravi. Try the delay loop. If it works, I think it's an appropriate fix for now as long as you put a descriptive comment block above it explaining exactly why it's there. And put the info in the CR, too."

"I'll make the change and load it into these mixers to test it."

Li Mei offered to check the code and help with the reload and testing. He was grateful for the help, but more so for the camaraderie.

Josie finished up the code review documentation for the final MixItMaster software and sent an email to the team thanking them for their hard work helping her nail the sneaky display problem. She made a point to thank Ravi publicly in the email for his ideas that helped them crack the second bug; he'd made a real turnaround the last few weeks and she hoped it would last.

In retrospect, she was glad she hadn't said anything to him or Oscar after the blowup.

Hearing a tap on the wall, she turned to see Li Mei poking her head through the doorway, and motioned her in with a wave. "Just finishing an email about the MixItMaster. It's off to System Test again." Josie continued to type.

"That was a neat bug, with all the timing problems, " Li Mei said. "Each of the symptoms pointed to something else, hiding the real problem."

"'Great battles kick up a lot of dirt. Obscures the battlefields so the generals can't see what's going on,'" Josie quoted. [1] "We had to clear out a lot of dust before nailing stack as the culprit."

"I never knew what a stack was before this. I didn't realize I had to pay attention to it."

Josie hit the send button with a flourish and spun her chair around to face her. "If the architects have considered the RAM requirements appropriately, you usually don't need to after the system is initially designed. But put it on your list of things to check when things get screwy like this."

"I did already," she said.

An idea occurred to Josie. "I asked Ravi to prepare a presentation on these display bugs for this month's Lunch and Learn. Maybe we should put you on the schedule to present your debugging list. What do you think?"

"Me?" Li Mei squeaked, eyes wide, and thumped herself in the chest with an open palm. "You want *me* to get in front of everyone and talk for 45 minutes about bugs?"

"Not just about bugs, but how to find them." Josie was already warming to the idea. "You could talk about how some really catastrophic failures can be caused by pretty simple bugs, and then give examples that you have seen since you started here. Start by presenting the symptoms of each bug, and then show how to use your list to identify the root cause."

Li Mei considered, still looking extremely nervous at the idea. "But what if people think it's dumb?"

"Then I'll be there to tell them how Ravi and I each used your list to identify the display problems with the MixItMaster. And how Oscar asked for your list to successfully demonstrate the Austin Monitor during the Telecom Trade Show." Josie tried to be convincing. "Everyone on this team has used your list to successfully zap at least one bug - isn't that a great endorsement?"

"Well," Li Mei's resistance was crumbling, so Josie decided to go for the kill.

"How about this, then. Performance reviews are coming up soon, and doing Lunch and Learns is something extra you can put on your evaluation to show initiative and teamwork. That stuff looks good when they are figuring out salary increases, trust me."

Josie nodded knowingly at her, watching the internal war play over Li Mei's face. Li Mei's response puzzled her; she was surprised that Li Mei was nervous about the informal gatherings in the training room over lunch, especially since she showed no fear talking at status meetings or challenging others about technical details in the lab. She'd been even more bold at customer sites.

Li Mei appeared to come to a decision.

"Okay, I'll do it. But not for at least 2 months," she added quickly, " so I can adequately prepare."

Josie chuckled and slapped her hand on the desk. "Excellent!"

"Where's Ravi?" Oscar asked.

"He's coming. I think he's playing a joke on Eduardo," Li Mei answered.

Oscar had called a Required Team Meeting at Molly's and the suspense was killing her. Josie checked her voice mail as she watched Li Mei hopping in place, her backpack bouncing on the small of her back. Shortly Ravi appeared around the corner, walking quickly toward them as they stood waiting at the employee entrance. He had a mischievous look on his face and she wondered if Eduardo would come screaming around the corner after him.

"I'll drive," he called. "I just want to get out of here fast!"

They found their way uneventfully to the corner booth at Molly's, everyone chuckling over the latest salvo while hoping to stay out of the crossfire. Maria delivered a round of drinks and appetizers at Oscar's direction and the chatter slowly died down as curious anticipation took over.

Li Mei nervously straightened her sweater and Ravi flipped a fork over his fingers as he looked around the table. Oscar took a deep breath and began.

"I knew I could get everyone here if I said we were leaving work an hour early."

Josie laughed with everyone else.

"Because I wanted to share some news that will be announced tomorrow. Hudson Technologies is acquiring another company and expanding into the Far East." He paused to gauge their reactions; everyone had gone on yellow alert. "First things first - no one is getting fired, and our team stays together."

Josie breathed a sigh of relief. She'd been instantly worried that the recent improvement in team dynamics would be inadvertently rewarded with a team breakup.

"Instead, we're getting bigger. Hudson's been trying to jump on the bleeding-edge technology bandwagon with some larger projects. I've been doing technical feasibility analyses for Jack Katalan."

Josie whistled low. It's not often the vice president of R&D taps underlings for special projects. No wonder Oscar's been keeping odd hours lately! She wanted to know the specifics and why Jack had gone directly to Oscar. And, she realized shortly after, why Randy wasn't in the loop.

Oscar continued, "Jack has some great ideas how we can leverage architecture and design brainpower in the US and couple it with implementation and manufacturing overseas. Everything is global now, and if we don't react, we're dead. He's embracing it with an eye to directly attacking problem areas like effective virtual collaborations and cross-cultural interactions."

"What does that mean for us?" Josie asked. "Will there be relocations?"

"No, but you'll probably get to travel overseas." Oscar nodded around the table to include Li Mei and Ravi. "Our first project is not finalized, but we might have a sister team in the Far East. We are also going to hire two or three more engineers for our team and you guys will all participate in the selection process."

Oscar looked at each of them in turn. "As we make these changes, I want you to spend your valuable time reviewing resumes and making good decisions on new hires. We can't afford toxic personalities, self-proclaimed super stars, or prima donnas. Jack recognized that we've got a good, effective team and he wants to keep it that way."

He paused, softening his voice slightly. "And so do I. It's going to be a new day at Hudson Technologies, and I have a good feeling for the engineers - for everyone - going forward."

Silence fell around the table and Josie lifted her glass to take an awkward drink, letting the disjointed thoughts clatter about in her brain. He'd just shaken their world with news that could radically alter their daily lives, and then threw in some unexpected praise for the team's dynamics. A positive anxiety filled her stomach and she wondered what the future held. Jack was a savvy guy and Oscar clearly followed him. Josie knew she wouldn't be far behind.

She lowered her glass and saw her smile of anticipation and excitement reflected in Li Mei and Ravi's faces.

A new day indeed.

Additions to Li Mei's List of Debugging Secrets

Specific Symptoms and Bugs

- Any software changes to a hardware interface or external device should be verified against the timing requirements for the device.
- Boundary conditions can cause sporadic behavior. Look for loop counter limits, ranges in logicals, and parameters that change with state.

General Guidelines

- Sometimes hardware-dependent software is unavoidable; if so, document heavily.
- Use hypothesis to induce rare bugs to occur more frequently by providing the appropriate stressors (more/larger inputs, increased loading, larger memory allocations, faster timing, etc.).

Chapter Summary: The Case of the Wandering 'E' (Difficulty Level: Harder)

The fix for the shifted MixItMaster display causes a different display bug. A single bad character is occasionally displayed at a gap in the display driver's memory map, which corresponds to a boundary condition. Ravi suspects one nibble of display information is getting lost, and finds the previous change to flatten the code's calling structure causes the microcontroller to send data to the display driver too quickly.

The Problem Symptom(s):

- The instrument display message contained one garbage character, only in the ninth position.
- The problem occurred more often than the previous shifted display problem.
- The problem happened with all mixing profiles.
- The bug was never seen before.

Targeted Search:

Li Mei recited her list for sporadic bugs, reminding the team about boundary conditions, memory corruption, and timing problems.

- The ninth position on the display corresponded to the beginning of the second half of the display driver RAM area used to store characters.
- The code to set position was not changed, but the functionality had been placed inline to flatten the calling tree.

The Smoking Gun:

Data were sent to the display driver one 4-bit nibble at a time. Also, the correct character 'n' and the garbage character 'α' both share an identical nibble, '0xE'. An 0xE received at the wrong time could change an 'n' to an 'α'.

The Bug:

Inlining the code caused fewer instructions to be executed. Successive nibbles of information were sent to the display driver too quickly and some nibbles were lost.

The Debugging Method Used:

- Brainstorming root causes of sporadic symptoms, focusing on memory corruption.
- Exploring ideas as a group rather than working individually.
- Creating debugging code that caused the bug to happen more frequently.

The Fix:

Since the hardware busy line was not available, the team reintroduced the necessary delay between successive writes to the LCD display driver using an assembly-language loop.

Verifying the Fix:

- The team used the debugging code that exacerbated the problem to verify the fix.
- System Test repeated the test plan for the entire product, including extra testing on the display itself.

Lessons Learned:

- Interfacing to hardware requires attention to timing specifications of the control and data signals.
- Writing to RAM on external devices takes longer than writing to RAM on the onboard device.

Code Review:

The inline code is harder to read, but fixes the stack overflow problem. The new hard-coded delays are machine-dependent, which can cause problems if the software is ported to faster hardware. Coding "at the metal" is often done in assembly language, making self-documenting variable names and comment blocks vital.

What Caused the Real-World Bug? When contacted about the slight overcharge for boat parts using their web-page online order interface, the company reported that they'd just fixed some problems with the ordering system, and somehow injected another bug. At least shipping was free! [2]

References

[1] *House*, "Sleeping Dogs Lie," Episode no. 40, first broadcast April 18, 2006 by Fox. Directed by Greg Yaitanes and written by Sara Hess.

[2] Ganssle, J., Personal communications, August 1, 2006.

Li Mei's List of Debugging Secrets

L i Mei developed her List of Debugging Secrets while debugging her own mysteries, participating as a sounding board for her teammates as they struggled with their debugging challenges, and listening to everyone's Lessons Learned at the team's regular status meetings. She will undoubtedly add to the list following each new challenge she faces.

Her list opens with the Action Plan that Josie explained to her, with specific observations gathered over the course of the book. It contains several Truisms she experienced ("Resist the Urge!", "Think with your brain, not your debugger.") and guidelines for writing better code. Finally, it concludes with a list of symptoms and bugs gathered from the individual mysteries the team solved, with a reference back to the chapter where the item appeared.

Li Mei's List appears on the following pages. Use it as a reference for your own work, and as a starting point for your *own* list of hard-earned Debugging Secrets.

Li Mei's List of Debugging Secrets

Action Plan for Solving Problems

1.   **Gather the Facts** - *Learn all you can before diving in.*
2.   **Classify the Symptoms** - *Characterize the bug based on the facts.*
3.   **Brainstorm Root Causes** - *Identify what could cause each symptom.*
4.   **Understand the System** - *Target where to search based on possible root causes.*
5.   **Hypothesize, Test and Verify** - *Be logical and methodical - nail the bug!*

1.   **Gather the Facts**
     o   Interview problem reporter.
     o   Interview anyone who saw the system fail.
     o   Observe the system behavior - find out what is <u>normal</u> for the system.
     o   Isolate relevant facts about product, customer, environment, hardware/ software, materials used, priority, safety.
     o   Reproduce the problem if possible.
     o   Realize bug report descriptions can be misleading.
     o   Be wary of <u>assumptions</u> when gathering facts - identify inputs and symptoms only. Draw no conclusions yet.

2.   **Classify the Symptoms**
     o   <u>One-time repeatable events</u> - occurrence has a pattern (e.g., only start up, different behavior first time through function or feature).
     o   <u>Periodic events</u> - regularly repeating or occurs every time (e.g., tied to timer, interrupt, repeated calls to function/feature, HW or SW heartbeat).
     o   <u>Sporadic events</u> - seemingly random failures (e.g., boundary condition violations, parameters that change with state, loop counter limits, ranges in logicals, unexpected input/output conditions, unhandled error conditions, faulty logic, hardware, timing, memory corruption, performance issues).

3.   **Brainstorm (Initial Root Causes, and When You're "Stuck")**
     o   Dream what could cause each symptom. The bug's location in the code can sometimes be determined before looking at the software.
     o   Identify the set of inputs that causes unacceptable behavior, then find what must be changed to make the behavior acceptable.
     o   Create a truth table to classify inputs (user, hardware, software, configuration, etc.) and resulting behaviors.
     o   Patiently watch the system's behavior.
     o   Periodically stop and summarize findings.

o Find a sounding board (doesn't have to be a live person!) - talk through ideas to quickly identify good ideas and discard bad ones.
o Consult the gurus.
o Talk to internal groups (engineering, testing, marketing, etc.) and external groups (vendors, customers, beta testers, etc.).
o Go home. Or somewhere else. Let your brain chew at the problem while you're doing something else.

4. **Understand the System (hardware, software, mechanics)**
   o Focus on understanding main() first for overall program flow - don't get lost in the details.
   o Divide the code into logical chunks based on structure and flow.
   o Use visual aids (flowcharts, graphs, function call trees) to show functional elements (blocks) and program control logic (connectors). This reveals what the program does, reveals structure and testing, and identifies missing logical and functional elements.
   o Play Computer. Doggedly step through the code line-by-line because sequential logic performed by a computer does not always match human assumptions. (Trace assembly language noting the contents of registers and stack pointer.)
   o Debugging tools allow simple timing characterizations without stopping the program execution.
   o When reverse-engineering code, check off functions (e.g., know where you've been, identify unused functions).
   o Remember sometimes the comments are wrong.

5. **Hypothesize, Test and Verify**
   o Decide exactly what information you would like to get from the embedded system, and choose the best tool accordingly.
   o Consider nontraditional and low tech debugging methods (e.g., auditory, pin wiggling).
   o Simple methods like pin wiggling can be powerful and unobtrusive.
   o Auditory cues - sense of hearing - can discern fine differences in tone and rhythm, and can also be used as a heartbeat or a code coverage flag.
   o Use black-box testing to fully explore behavior without looking inside the device/software.
   o Hypothesize what should happen in order to choose which variables to watch.

- Apply stressors to induce rare bugs to occur more frequently (more/larger inputs, increased loading, larger memory allocations, faster timing, etc.).
- Use breakpoints configured as watch points to check when values of variables and memory locations change without stopping code execution until the exact conditions you specify are met.
- Use patterned memory (e.g., DEADBEEF, 0x55) to verify memory operations and to identify memory overruns.
- Test bug fix with methods originally used to induce the bug in the first place.

Truisms

- "Resist the Urge!" to jump blindly into the software listings.
- Think with your brain, not your debugger.
- Just because it compiles doesn't mean it works.
- Be clever.
- Don't worry about what other people think. Just worry about getting it fixed.
- Don't make assumptions about how something was implemented (e.g., in hardware versus in software). Look at the evidence and the documentation.
- Commercial tools can contain bugs; if a tool does something strange, suspect the tool.
- When you feel overwhelmed, take things one step at a time.
- Randomly changing lines of code is not an effective debugging technique!

Making Better Software

- Use programming elements that are appropriate for the function (e.g., switch for unrelated discrete items and if-statements for continuous variables).
- Make code self-documenting with descriptive names, #defines, and comments.
- Consistent tab spacing and white space make code more readable.
- Sometimes hardware-dependent software is unavoidable; if so, document heavily.
- Coding standards are a good source of bug types and causes, and also to search for possible fixes!

Mystery-Specific Symptoms and Bugs
- Variables
  - Initialize all variables. Don't assume the compiler will do it for you. (Ch 2, 3)
  - Avoid using hard-coded numbers. If a #define or enum is available, use it. If not, create one with a descriptive name only if you are SURE what it does. (Ch 1, 3, 4, 5, 8, 9)
  - Make sure unsigned variables will not be used to store signed quantities. (Ch 4)
  - Suspicious indexes into arrays often cause off-by-one or rollover errors. (Ch 1, 2)
- Performance
  - Reduce overhead when transmitting data to conserve battery power. (Ch 7)
  - Sending more than one data sample at a time and/or compressing data in the message payload can radically increase the life of battery-operated devices. (Ch 7)
  - Remember to turn off debugging code before shipment! (Ch 6)
  - Test scripts should use existing functions to duplicate normal device operation. If a test-only function must be created, document what it does and why it is there. (Ch 1)
  - Inlining code can improve processing speed. (Ch 8, 9)
- Hardware and Timing
  - Document any underlying hardware assumptions. (Ch 3, 8, 9)
  - Any software changes to a hardware interface or external device should be verified against the timing requirements for the device. (Ch 2, 3, 9)
  - Writing to RAM on external devices takes longer than writing to RAM on the onboard device. From software, hardware can look like a variable - don't treat it like one. (Ch 9)
  - When controlling motors, use absolute (rather than relative) motor position references for known, repeating activities. Send the motor home regularly. (Ch 5)
  - Don't assume the motor does what you tell it. Check it. (Ch 5)
  - For proportional errors (twice as fast, off by 3, etc.) check if data are periodically missed, two entities expect data at different rates, or a configuration setting is bad. (Ch 7)

- ISRs
  - Make sure only time-critical functions are included in Interrupt Service Routines. Verify the duty cycle. (Ch 3, 9)
  - Split functions or use flags to remove noncritical code. (Ch 3, 7)
  - Only turn interrupts off for a very short time; otherwise, latencies increase. (Ch 4, 8)
- Logic
  - When using flags for signaling, ensure the entities check the state first to avoid missing information. (Ch 7)
  - RTOSes can allow priority inversion if tasks of different priority levels can access the same common system resources (e.g., memory). (Ch 6)
  - Check logic to ensure counting variables are bounded ("<=" versus "=="). (Ch 5, 9)
- Memory and Stack
  - When something works for a while before breaking, suspect memory problems like boundary condition violations. (Ch 1, 2, 8, 9)
  - Ensure variables stored in two locations (like RAM and FLASH) are synchronized. (Ch 1)
  - Perform stack analysis early and often to understand memory usage. When stack size is limited, or when user variables are allowed to share RAM with the stack, stack overflow can cause catastrophic and unpredictable results. (Ch 8)
  - Suspect stack corruption when problems manifest in deeply nested functions or when a lot of data is passed between functions. (Ch 8)

Index

Printed and bound by CPI Group (UK) Ltd, Croydon, CR0 4YY

03/10/2024

01040336-0003